Electronic Surveillance and Wireless Network Hacking

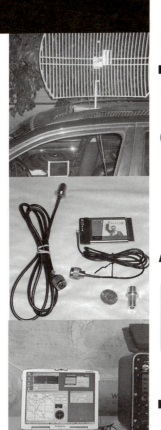

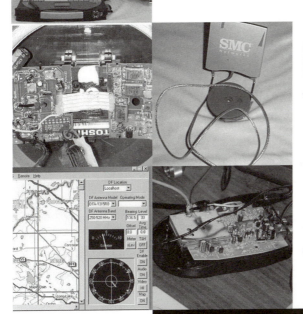

Protecting Your Privacy in a Wired and Wi-Fi World

Paladin Press
Boulder, Colorado

M.L. Shannon and Steve Uhrig

Also by M.L. Shannon:

The Bug Book: Everything You Ever Wanted to Know about Electronic Eavesdropping . . . But Were Afraid to Ask

Digital Privacy: A Guide to Computer Security

Don't Bug Me: The Latest High-Tech Spy Methods

The Phone Book: The Latest High-Tech Techniques and Equipment for Preventing Electronic Eavesdropping, Recording Phone Calls, Ending Harassing Calls, and Stopping Toll Fraud

Electronic Surveillance and Wireless Network Hacking: Protecting Your Privacy in a Wired and Wi-Fi World
by M.L. Shannon and Steve Uhrig

Copyright © 2005 by M.L. Shannon and Steve Uhrig

ISBN 10: 1-58160-475-0
ISBN 13: 978-1-58160-475-7
Printed in the United States of America

Published by Paladin Press, a division of
Paladin Enterprises, Inc.
Gunbarrel Tech Center
7077 Winchester Circle
Boulder, Colorado 80301 USA
+1.303.443.7250

Direct inquiries and/or orders to the above address.

PALADIN, PALADIN PRESS, and the "horse head" design are trademarks belonging to Paladin Enterprises and registered in United States Patent and Trademark Office.

All rights reserved. Except for use in a review, no portion of this book may be reproduced in any form without the express written permission of the publisher.

Neither the author nor the publisher assumes any responsibility for the use or misuse of information contained in this book.

Visit our Web site at: www.paladin-press.com

DECENCY, SECURITY, AND LIBERTY ALIKE DEMAND THAT government officials shall be subjected to the same rules of conduct that are commands to the citizen. In a government of laws, existence of the government will be imperiled if it fails to observe the law scrupulously. Our government is the potent, the omnipresent teacher. For good or for ill, it teaches the whole people by its example. Crime is contagious. If the government becomes a lawbreaker, it breeds contempt for law; it invites every man to become a law unto himself; it invites anarchy. To declare that in the administration of the criminal law the end justifies the means—to declare that the government may commit crimes in order to secure the conviction of a private criminal—would bring terrible retribution. Against that pernicious doctrine this court should resolutely set its face.

—Supreme Court Justice Louis Brandeis
in his dissenting opinion in the Olmstead case.
OLMSTEAD v. US, 277 US 438 (1928)

Contents

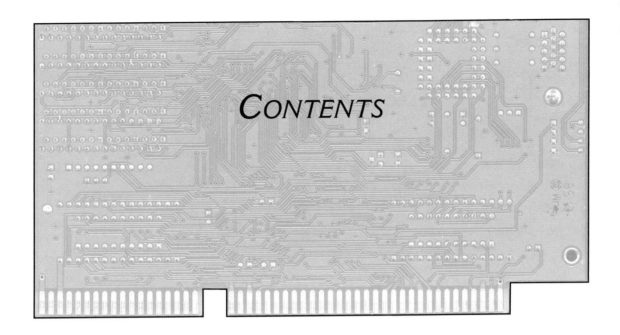

Acknowledgments -- ix
    Photo Credits -- *ix*
    Producing This Book -- *x*
The Internet Edition of *Electronic Surveillance and Wireless Network Hacking* ------------------ xi
Disclaimer -- xiii

Part One:
Privacy

A Reasonable Expectation of Privacy -- 3
    Audio --- *3*
    Video --- *3*
    Telephones --- *4*
    The Internet -- *4*
    The Game is Electronic Tag and You Are It ------------------------- *4*
RFID: Radio Frequency Identification Devices ------------------------------- 9
    How They Work: Batteries Not Included ---------------------------- *10*
    Collect and Trade, Mix and Match --------------------------------- *11*
    If You Won't Join 'Em, Jam 'Em ----------------------------------- *11*
Broadband Over Powerline -- 13
    Yes, Big Brother is Still Listening ------------------------------ *13*
    How It Works -- *14*
    Hacking BPL --- *14*
    The BPL Party Line? --- *14*
    The (BPL) Colony? --- *15*
    RFI: Radio Frequency Interference -------------------------------- *15*
    Unplugging Homeplug --- *15*

Part Two:
ELECTRONIC SURVEILLANCE

Audio Surveillance — 19
- *Cell Phones* — 20
- *Digital Recorders* — 20
- *Lasers* — 20
- *The Infinity Transmitter* — 20
- *What a Spy Might Hear* — 21

Video Surveillance — 23
- *Wireless Video Cameras* — 24
- *Radio Frequency Surveillance: Bugs* — 25
- *What Is New* — 26
- *Spread Spectrum* — 27

Wireless Telephones and Privacy — 29
Radiation Warning — 33
Wave Propagation: Understanding Radio Frequency Signals — 35
Wiretapping — 37
- *How Do I Search for a Phone Tap?* — 39
- *What is a Pen Register?* — 41
- *Answering Machines* — 41
- *The Drop-Out Relay* — 41
- *The "Tele-Phone"* — 42

Other Surveillance Methods — 45
- *Vehicle Tracking* — 45
- *Natural Audio Paths* — 46
- *Flickering Lamps* — 46
- *RF Flooding* — 46
- *Subcarrier Transmitters* — 47

Spying in Real Life: Surveillance Installations — 49

Part Three:
ELECTRONIC COUNTERMEASURES

Introduction to TSCM — 53
Seeing Red with Kevin Murray: Infrared Thermal Imaging to Locate Surveillance Devices — 57
- *Science Catches Up to Our Dreams* — 58
- *Advantages of Thermal Emissions Spectrum Analysis* — 58
- *About the Author* — 58

TSCM 101: So You Want to Be a Sweep Technician, Eh? — 61
The Bushwhacker™ from CSE Associates — 63
Well, It Looked Like a Bug ... — 65
Well, It Is a Bug! — 69

Part Four:
THE INTERNET

The Information SuperSpyway — 75
Cyberterrorists and the Law — 77
The "Cyber-Street Survival" Series — 79
    Part 1: Getting Started — 79
    Part 2: Spam: Just Say Delete — 79
    Part 3: Without a Trace — 80
    Part 4: Security and Other Things — 80
    Part 5: Hackers — 80
    Part 6: Internet Tools — 80
Internet 101 — 81
    Getting Started — 82
    Firewalls — 82
    Log Files — 83
    Forms — 84
    Banner Ads — 84
    Ad Blockers and Other Privacy Applications — 84
    Cookies — 85
    Dog Cookies — 85
    Proxy Servers — 85
Identity Theft and How to Prevent It — 87
What Is Hacking? What Is a Hacker? — 89
    Black Hats and White Hats — 90
Internet Tools — 91
    net.demon — 91
    IP Address Allocation — 92
    Back to Configuring net.demon — 92
    IP — 92
    Ping — 93
    Traceroute — 93
    VisualRoute — 94
    WHOIS (Who Is?) — 94
    Stupid URL Tricks — 95
    Finger — 95
    WWW — 95
    Information — 95
    Now, Back to Configuring net.demon — 96
    E-mail Verify — 96
Secret Computer Codes: Data Encryption — 97
    What About the Data Encryption Standard? — 98
    What is PGP? — 99
    What Exactly Is RSA? — 100
    What is Steganography? — 101

Part Five:
IT'S A WIRELESS WORLD

- Introduction to Wireless Networking — 105
- Getting Started in Wireless Networking — 109
  - *Do Try This At Home* — 109
  - *Home Network* — 111
- Wireless Network Monitoring and the Law — 113
- Sniffers and Other Wireless Applications — 115
  - *WinPcap* — 116
  - *NetStumbler* — 116
  - *CommView Wireless* — 118
  - *Ethereal* — 123
  - *Auditor* — 123
  - *Knoppix* — 124
  - *Kismet* — 126
  - *Wireless Cards* — 127
- SSIDs: Wireless Network AP Names for Fun but Not Profit — 131
  - *A Better Way* — 132
- Product Review: Wi-Fi Seeker — 133
  - *Batteries Included* — 133
  - *Field Test* — 134
- Wired Equivalent Privacy — 135
- Intruder Alert: Someone Is Using My AP! — 137
  - *Spoofing the MAC* — 138
- A-Hacking We Will Go — 141
  - *Programming the Router* — 142
- War Driving — 145
  - *Prevention: What Can You Do?* — 146
  - *War Chalking* — 147
- A Site Survey — 149
  - *The Survey Report* — 150
- Wireless Networking Antennas — 153
  - *Building Wi-Fi Antennas* — 154
- Wireless Networking FAQ: Questions and (Some) Answers — 155

Part Six:
REAL SPIES, REEL SPIES, AND REAL "NUTZ"

- The Fascinating World of Minox — 161
- On Location: Filming *Enemy of the State* — 165
- The "Nutz" File — 167

- The Last Word — 175
- Appendix A: Sources — 179
- Appendix B: Suggested Reading — 181
- Appendix C: Electronics 101 — 183
- Glossary — 185
- About the Authors — 189

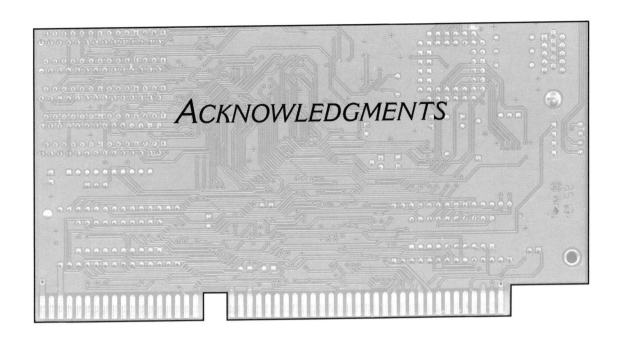

ACKNOWLEDGMENTS

A NUMBER OF PEOPLE HAVE MADE CONTRIBUTIONS TO THIS WORK, FOR which I am most grateful. In no particular order:

Coauthor Steve Uhrig for a great deal of what is contained here, especially photos from the filming of *Enemy of the State* and the Minox chapter; Kevin Murray of Murray Associates, for photos of the thermal imaging system; Ozzie Eans at CSE Associates, for permission to use pictures of the Bushwhacker direction-finding system; Rick Hofmann of Microsearch LLC, a professional countermeasures team, for the chapter, "It is a Bug!" and Tamos Software, producers of CommView for wireless.

Also, "Max," who lives in Switzerland, the creator of the Auditor bootable Linux CD; and Matthew Schneider, who wrote net.demon, the excellent suite of IP tools.

And, the producers of Network Stumbler; people at SourceForge, for permission to publish screen shots; "EvilMofo," who is best known for being a BSD wizard and for his stack of burned-up motherboards. (Also for his spotless and perfectly organized work area and careful driving habits.) Evil provided pictures from the fourth WorldWide WarDriving contest.

And several other people, including hackers and geeks by any other name, who don't want to be named.

PHOTO CREDITS

M.L. Shannon's photos were taken with a Nikon D-100 digital SLR camera and a small Sony Cyber-Shot.

Some photos by Steve Uhrig were taken with an Olympus XA 35mm but most were taken with a Minox.

In "On Location: Filming *Enemy of the State*," the pictures were taken on location by Steve or by the production's "stills" photographer. See our Web site at http://www.fusionsites.com/dbm2 for all of them.

Some pictures at *Electronic Surveillance and Wireless Network Hacking* on the Web were taken by various geeks at San Francisco 2600 hackers meetings.

PRODUCING THIS BOOK

This book was written on a Pentium III with a Sony Trinitron 17-inch monitor using, mostly, the Easy Office word processor, then into Word for the publisher. It was also imported into NetObjects Fusion MX and formatted as a Web site, which was uploaded for various persons to view and comment on, including some of those who contributed to the manuscript. Some graphic work was done in PhotoShop, PaintShop Pro, and an ancient version of Corel Draw.

The Internet connection is ADSL through a Siemens SS-2624 router/firewall and wireless AP networked to a Compaq Presario Notebook Pentium III, and a Pentium II desktop running Linux Red Hat 8.0 professional. While I was online to the Internet, whether I was using dialup, DSL on one machine that was CAT-5 cable to the router, or one of the other computers that used wireless from the DSL router, I had an early version of Zone Alarm and Win Patrol running at all times.

Wireless research was done using Senao, Orinoco Gold, Proxim, LinkSys WPC-11 and LinkSys WPC-55AG PCMCIA cards, LinkSys WMP-11 PCI card and several antennas. All are pictured in the appropriate chapters. The one that looks like a black tube (it *is* a black tube) is a 14-dB Yagi from Antenex. The large one is a 24-dB grid. I lucked out and found a tripod at Goodwill that had a flat aluminum top and had the antenna bracket welded in place so that it could be precisely aimed.

Mapping for the "Wireless Survey" chapter was done with various shareware applications and a Garmin GPS-72 global positioning receiver.

Software used includes, for Windows: NetStumbler, CommView Wireless, Capsa, WinPcap, Ethereal, and Packetyzer. For Linux, Knoppix and Auditor were used, each of which contains dozens of utilities including Kismet.

The Internet Edition of Electronic Surveillance and Wireless Network Hacking

THIS BOOK HAS LOTS OF PHOTOS, AS YOU CAN SEE; HOWEVER, BLACK AND white doesn't do them justice, and the screen shots are too small to read the details, so I have decided to build a section on my Web site where you can see them in color, as well as read a little additional information about when and where they were taken.

There will also be some shots for which there was not enough space in the book; some that could not be used for legal reasons, including a few taken on location during filming of *Enemy of the State*; a few from San Francisco 2600 meetings; and perhaps some new ones from New Zealand.

The URL is http://www.fusionsites.com/dbm2. (Why dbm2? *Don't Bug Me II* was my working title for this book—a sequel to the original *Don't Bug Me* that I wrote for Paladin Press.) An exclusive: on this site you can hear the actual phone conversation of a near hysterical person who called 911 to report being under electronic surveillance. This lengthy .wav file is complete except that the names have been dubbed out.

I also plan to set up a private chat room, if there is enough interest, and I can usually be found on the IRC channel #SF2600 on EF Net.

Additional pictures of interest are on Steve's site, http://www.swssec.com. You can look, but some products at SWS Security are available only to law enforcement agencies.

Please be aware that all pictures are copyrighted.

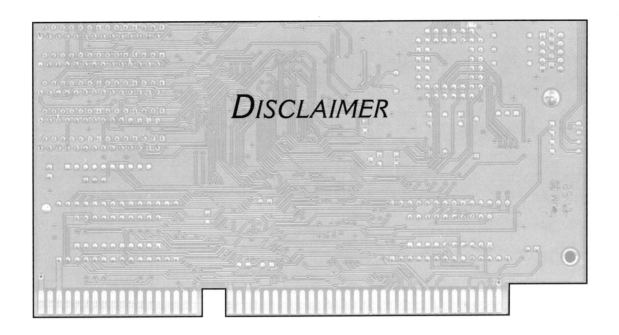

DISCLAIMER

THIS IS NOT AN ACADEMIC BOOK—IT IS A HANDS-ON GUIDE WITH WHICH THE average person who has little or no knowledge of electronics can protect himself and his family and business from unlawful intrusion by those who would spy upon him. While it contains details on the methodology of electronic spying, hacking, and wireless network monitoring, this is included only so that readers may understand how these things work so that they can learn to deal with them. Again, it is intended only as a comprehensive self-defense manual.

Some of what is in this book could be used for unlawful purposes. It is not my intention to advocate breaking any law, local or federal. You are strongly advised to consult an attorney before you do anything that could possibly be in violation of existing statutes. Keep in mind that, as stated later in this book, while existing statutes appear for the most part to remain unchanged, we now have Homeland Security and Total Information Awareness and whatever else the federal government will have enacted by the time this book is published. And, unfortunately, the government is not forthcoming about these laws. We The People don't know—are not told—what the government is up to, what is or is not legal.

Neither the author, the publisher, nor the seller assumes, or will assume, any responsibility for the use or misuse of the information contained herein, nor will the author, publisher, or seller be responsible for the consequences thereof, either civil or criminal.

The computer system on which this book was written, produced, and typeset is linked to, and part of, the electronic mail system, including but not limited to the Internet. Some of the material on this computer system, including all future editions and revisions, is being prepared, or was prepared for public dissemination and is therefore "work product material" protected under the First Amendment Privacy Protection Act of 1980 (USC 42, Section 2000aa). Violation of this statute by law enforcement agents may result in a civil suit as provided under Section

2000aa-6. Agents in some states may not be protected from personal civil liability if they violate this statute.

In the chapters on wireless networking, all of the exercises in which actual data was intercepted were done on my own private network between two of my own computers. The SSIDs (service set identifier) shown in the various screen captures do not contain data—only the SSID, signal strength, etc. In other words, I did not hack into anyone's network in order to complete this book. It wasn't necessary: I hacked myself!

Part One

PRIVACY

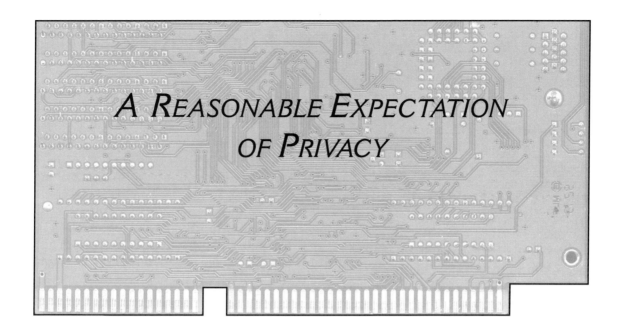

A Reasonable Expectation of Privacy

WHAT IS A REASONABLE EXPECTATION OF PRIVACY? WHAT IT COMES DOWN to, in the years since the terrorist attacks of September 11, 2001, and Total Information Awareness and Homeland Security, Carnivore, Echelon, and Magic Lantern ... is that with few exceptions, you don't have any privacy. Here's where we stand legally on several different forms of surveillance.

AUDIO

Technically, no one is allowed to record your in-person or telephone conversations unless they comply with the law. Federal law, unless it has recently changed, requires what is known as single-party consent, which means that one party to the conversation must be aware that it is being taped. So, if you call Joe and want to capture what he has to say, you can do so because you are a (single) party to the conversation. The reasoning here is that it is unlawful for someone who is not a party to the conversation to record it. In other words, an eavesdropper or a wiretapper.

The all-party law requires that everyone who is part of the conversation is made aware of the recording by the person doing it. In person this means verbally so advising them and with telephone conversations it means playing a beep tone on the line every 15 seconds so everyone can hear it. Some phones, like the Panasonic Auto-Logic, have this capability built in. For more information, see the Web site of the Reporters Committee for Freedom of the Press at http://www.rcfp.org/taping/.

VIDEO

There are some laws regulating video surveillance in general, but there are no "consent" regulations that I know of. As I have mentioned before, and will again, it is a good idea to talk to a lawyer whenever there is any doubt as to what is and is not legal. Remember about legislation the public has not been made aware of.

Some of these laws concern invasion of privacy. There are laws forbidding cameras in restrooms, locker rooms, and changing rooms in department stores, but like any other law, they get abused. If you read newspapers you will have seen many articles about how this assumed right to privacy is violated on a regular basis by those who have no regard for the law.

TELEPHONES

This can be summed up in few words: When you talk on the phone, assume that something is listening. Something? Yes, computers are intercepting your calls—all of your calls—looking for certain words and phrases. If detected, then a live human will most definitely be listening.

THE INTERNET

In the 20 or so years before the Internet became available to the general public, privacy was less of an issue. True, there were hackers who knew Unix well enough to get into many computers (including some belonging to the military), copy files, and read e-mail; *The Cuckoo's Egg* (see Appendix B) is the fascinating true story of how one such hacker was caught. But this was before millions of people started sending personal e-mail, credit-card numbers for online purchases, and all manner of other personal information about jobs, health issues, and whatever else. It was before the use of some Internet programs that were, and still are, so full of weaknesses that they were easy to exploit; before careless companies left much of this personal information lying around where practically anyone could access and copy it; before identity theft was so prevalent.

That was then. This is now.

Again: Talk to a lawyer, and if you want to do some research yourself, start with 18USC 2511/2512/2513. That's the United States Code, Federal statute law, available in any law library.

The next section is an example of what might happen, the surveillance and privacy-invading techniques that might be used against a "person of interest," on a not so ordinary day.

THE GAME IS ELECTRONIC TAG AND YOU ARE IT

The alarm goes off at 6:30 A.M. and you stumble out of bed and into the bathroom, turning the light on. It's a big day; you have an important meeting at work and later, a flight to a nearby city to present your proposal to a marketing company. A few minutes later the phone rings.

"Good morning, this is your wake-up call" emanates from the answering machine and you remember that you arranged for the call just in case the alarm failed or you rolled over and swatted it off and went back to sleep.

You shower and get ready for work, check your briefcase for documents—passport, driver's license, and other essentials—get in your car, and head for the office. You sometimes forget that, in order to get a lower insurance premium, you agreed to have a "flight recorder" device installed in the vehicle. Since you forgot to call the local deli to have a take-out order ready, you are running a little behind so you punch the gas and go over the speed limit by 10 mph. This causes your wireless PDA to beep, telling you that your insurer has detected this infraction of the law and you are warned that this could cause an increase in your rate.

On the way you call the office on your cell phone to ask the receptionist if a certain person has arrived. You realize you are a little low on gas so you stop at a neighborhood station to fill up using one of those swipe cards and, finally, enter the underground parking area using yet another plastic card.

The meeting begins and ends, boring as usual, then you go back to your cubicle, where your terminal is part of the company wireless network, to take care of your e-mail. Then you take a break for lunch. The company cafeteria being what it is, you and a few others head out to a local restaurant and have a few drinks with your food. On the way back you stop at an ATM machine to get some cash and from there, it's back to the office where you gather up your material and, leaving your car in the garage, take a cab to the airport. On the way, your pager starts vibrating and you read the message from your secretary—you forgot your file on the Phydeaux Corporation. Fortunately, you won't really need it, so no big deal.

You get your boarding pass, and after the usual hassle of waiting in a long line trying to get through security, you go through the skyway onto the jet, which takes off and lands without incident.

The presentation is delayed because a key participant isn't there yet (plane delayed or whatever) and doesn't get started till midnight and ends at 4 A.M. Fortunately, there is a late flight and you are able to get back about 6 A.M. You grab a cab, and on the way home, you remember that there is a dinner party that evening. You stop at an all-night supermarket to get several bottles of decent Chablis using your "smart" card (those cards issued by some merchants and supermarkets that record your buying preferences), and finally head home where you get a few hours sleep.

This is a typical day in the life of some busy executives, and while most people's lives aren't always this hectic, there are similarities in the meetings, cab rides, ATMs, etc.

Now suppose, for whatever reason, a federal agency (not necessarily American), was interested in you. In what you do every day. What might they have learned about you on this particular day?

Thousands of miles away, a company you have never heard of has a system that makes a record of every call you make or receive. They know you had a wake-up call, something you don't ordinarily do. So, something important might be happening. They have a closer look.

Your plastic card at the gas station makes a permanent record of your purchase, which is uploaded by satellite to a data-processing location used by the gas station. (Ever notice those small dish antennas on the roofs of the offices?) Now, a profile can be made of your gas purchases and how much you use an average week, and should you vary from this significantly, someone knows you are going somewhere you usually do not. By digging further they know when you go on vacation (if you ever have the time) and the usual trips you make. More personal data being stored somewhere.

Your building management company maintains a record of when you enter and leave the parking area, which you may have realized, and you already know that all credit card purchases are instantly available to the feds if they have a "flag" on you, as are all your travel arrangements, such as which flight to where, what time, etc.

You used a credit card to pay for lunch and it made a permanent record of the alcohol purchase. Even though it might have been the other guys who had the martinis, you paid the bill and it is on your record. (Drinks during the day.)

You won't know it, but a tiny camera might be watching you every second you are on the aircraft. A company called Skyway Communications is allegedly making an air-to-ground network on which these cameras operate and installing them on the aircraft of certain commercial airlines.[1]

When you made the wine purchase at the supermarket, a permanent record was made and placed in a storage area, available to the government without a subpoena, without a warrant. You bought alcohol early in the morning; it makes no difference that it is for a dinner party, the record stands. Included is your name, the credit card number used to pay, your home address, and a detailed profile of what you buy, what brands, etc.

Naturally the ATM withdrawal—which bank at which address—is part of your records, but something else to consider: As long as your cell phone is turned on (does anyone ever turn them off?), the computer system that operates the phones has a constant record of where you are. With micro-cell technology, they know your location within about 15 feet at any time. And they can maintain a permanent record.

Now, suppose you paid cash at a particular store—didn't use a credit or "smart" card. They know which stores you usually patronize and, with that info along with the cell-phone location record and the time you spent within a given microcell, they have a good idea where you were shopping.

The last time you flew, a tag containing an RFID (radio frequency identification device) microchip was placed on your suitcases at the airport. When you pass certain locations, these chips are activated by scanners that record your movements. These chips are so small they may be placed unnoticed in a sport coat you bought, in a shirt you took to the cleaners, in virtually anything. Eventually, they will be in your cash—the $10s and $20s in your wallet.

The message on your pager might have been intercepted by experts, but it is also possible that someone who knows how to build a simple decoder and interface it to a computer might also have read about the forgotten Phydeaux Corporation file.

In a van, parked on the street near your office building, someone has set up a powerful directional antenna connected to a notebook computer that runs a number of wireless network detection and interception software applications. Whoever they are, they already know the SSID of your network and the MAC of your terminal, and their sophisticated system has captured the e-mail you send.

You may choose not to believe all this is really happening. It can and maybe it is. All these surveillance methods (and others) have been operational for some time. For example, in a statement titled "FCC Proposes Rules to Meet Technical Requirements of CALEA" [Communications Assistance for Law Enforcement Act] released on October 22, 1998, the Federal Communications Commission expressed its initial approval of FBI-proposed technical requirements that would enable law enforcement to determine the location of individuals using cellular telephones.

And the U.S. Department of Transportation is working on a plan that will track the location of every motor vehicle in the country and maintain a permanent record of where that vehicle has been. News items state that within a few years, all new vehicles made in the United States will have a black box device similar to the flight recorders in aircraft.

OK, you believe the technology exists but you still do not believe that these things are happening to you, and you are probably right. Unless a government agency or a private company that has the technology and the connections has a special interest in you, much of this surveillance isn't gonna happen. At least not directly. No one watching your every move.

But it *is* true that a permanent record is made (at least for as long as the phone company maintains it) of every call you make. And the same with your credit card transactions. And your supermarket purchases. And, again, this can be accessed by the feds without a warrant. Just for the asking.

Now, understanding the above, look for the signs of surveillance in your life. Are there people who seem to know things they are not supposed to know? Do they sometimes let something slip, say something and then try to cover it up or change the subject to draw your attention away from what they said, or started to say? Do some people always seem to know what you will do next and are they prepared for it? In a competitive business, are there other companies that repeatedly underbid you by a very

> SOMETIMES PEOPLE FIND IT A BIT of a hassle to keep track of all the cards they carry. But unknown to most Americans, within the next few years only one card will be necessary. It will have a microchip that contains everything there is to know about the holder and, like most Americans, these cardholders will think that this is so thoughtful of the government, making life easier for them. Will people ever learn? I doubt it.

small margin? Has your Aunt Martha apparently discovered the plot to the play *Arsenic and Old Lace* and asked her attorney to draw up a new will or a last-minute codicil, cutting you out completely?

Who are you, and what are you involved in that would cause someone to want to overhear your conversations? Are you an attorney working on a civil action or a corporate merger in which millions of dollars are at stake? Perhaps you're an advertising executive who is finalizing the plans of a massive ad campaign for a new product in an industry where there is a great deal of competition? Are you someone who is involved in (or who law enforcement believes is involved in) illegal activities such as drug dealing?

Do you have, for whatever reason, enemies who would like to "get the goods" on you? Are you in a relationship that you know is coming to an end, with considerations such as child custody or a property settlement?

If so, then there is always the possibility.

As to opportunity, whether or not someone could spy upon you, consider a few things. An analogy I use in *The Bug Book* is the possibility of bugging the White House. I could never bug the Oval Office. No way could I even get on the grounds. But Nixon did. Remember the 18 minutes of Watergate tape? Opportunity.

Your husband or wife, obviously, has access to your bedroom. So does anyone else if you leave your front door unlocked while you are both at work. If you have your home well secured (a sophisticated alarm system, Medeco locks on the doors, a German Shepherd in the backyard, and neighbors who look out for each other, etc.), then it is unlikely that anyone will be able to hide a bug inside your mattress.

Should you have reason to believe that someone wants to spy on you, perhaps this is something to think about.

Next, let's look at some surveillance technology.

ENDNOTE

1. Skyway Communications is SWYC on the NYSE. Last check their stock was 8 cents a share.

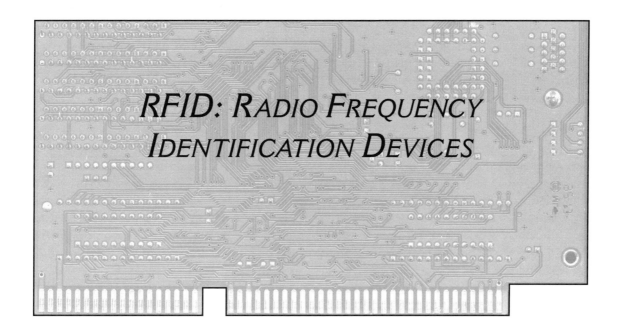

RFID: Radio Frequency Identification Devices

The RFID chips you read about in the last chapter are very real and a threat to your "right" to be let alone. There hasn't been much coverage in the mainstream media and you might not have known these chips even existed unless you read some of the alternative Web sites such as http://www.whatreallyhappened.com.

Businesses love them because they can be used for many things, such as inventory control. You buy a six-pack of Vernors Ginger Ale at the supermarket and when the scanner detects the chip, the item is removed from inventory in the store's computer. After a certain quantity of this soft drink is sold, the manager knows it is time to order more.

Tags attached to pallets of merchandise register when they pass the sensing scanner, and a computer not only makes an inventory adjustment but also directs the trucks to where the pallets are to be placed in a large warehouse. The uses are limitless—as are the abuses. They can, and are, being used to invade our privacy, although the manufacturers of these insidious chips naturally deny that this is true.

The chips can be imbedded into an employee's ID card to automatically generate a record of when he enters and leaves the building. Fine. Would eliminate the need to punch in and out at the time clock. But then they could also be used to show how often you go to the bathroom.

Apartment buildings could use RFIDs to keep track of when you are home or not home.

Attached to motor vehicles, they could eliminate waiting in line at toll booths (something that the average commuter crossing the Golden Gate Bridge in San Francisco would dearly love). The system would record your tag number and you would receive a bill at the end of the month. Great. But the sensors could be placed any number of other places and would likewise generate a record. Of everywhere you go. They could even be embedded in public transportation monthly passes to record how often you ride the cable cars or ferries.

Eventually, such a system would keep a record of everywhere you go. What buildings you enter, restaurants you patronize, theaters you frequent. The possibilities are endless, limitless.

Some people don't much like the idea that whenever their credit cards are used, a record of when, where, and for what is available, instantly, to any government agency without a warrant or even a subpoena, so they pay cash. But it won't be long before these chips will be embedded in said cash. The RFID numbers of the $10s and $20s that you get when you cash your paycheck at a bank will be recorded and placed in your account record. And when you spend them, more records are created. It could get to the point where you can't spend any money without Big Brother knowing the details.

Put the two together. The Gillette Company is supposedly placing RFID chips in their disposable razors. So, you buy a pack and pay for it with a $20 bill that also has a chip—the $20 that you got at the bank. And all of this gets electronically transferred to some massive government database and they have all the details of the transaction. Your name, what you bought, with the date, time, and place. Do you ever patronize businesses that you'd rather no one else knew about?

A company in Germany, Microtec, has produced washable RFID tags that can be sewn into garments, and it is only a matter of time until there are tags built into adhesive material that can quickly and permanently be attached to the clothes you take to the dry cleaners. Again, the possibilities are endless.

So, what exactly are RFID tags?

RFID tags are tiny microchips that contain unique information, similar to bar codes, only they hold more data. When one of these chips comes within range of a scanner, the scanner reads and stores the data in the chip. Now, depending on whom you believe, once the chips have been scanned they are supposed to be deactivated. So the one attached to your six-pack of Vernors might no longer function.

But what about tags used for collecting turnpike or bridge tolls? They can't be deactivated; they have to keep working indefinitely. When anyone tries to tell me that any chip is deactivated I don't believe it.

HOW THEY WORK: BATTERIES NOT INCLUDED

In some of the Internet media articles, it is implied that these chips are transmitters that constantly broadcast a signal at a certain frequency and that they can even be detected by orbiting satellites. This is not true.

The chips have no power source. Even the tiniest battery would be many times larger than the chip, which would limit their use in such items as currency. A battery would sooner or later go dead, and to change a battery would not be cost effective.

The RFID chips broadcast the data they contain only if they are within range of the scanning device. A radio signal is broadcast by the scanner, which activates the tag. I don't know exactly how these things work. One possibility is that the signal from the scanner causes a small voltage drop across an inductor. This voltage forward biases (powers) a transistor that oscillates and somehow radiates (sends) the data bits it contains. This signal is so weak that its range is measured in feet rather than hundreds of miles to orbiting satellites. Think about it. A cell phone, which does have a battery and a power output of a couple hundred milliwatts (tenths of a watt), can't reach (transmit to) a satellite directly; it also has to be relayed.

The actual range varies from a few inches to several feet for most tags, but there are others that work up to maybe 10 yards. These chips and sensors would be larger—which would limit where they could be used. Perhaps they would keep track of large objects, such as shipping containers.

So, as you have seen, while RFID chips have many useful purposes in industry, they can also be used to track you and maintain records of everything you buy and everywhere you go, building a profile of your lifestyle. And all of this information can, and probably will, end up in a federal government database.

This is the future. So what can you do about it?

Whenever possible and if it is not too inconvenient, avoid buying things that contain them. We will soon see a growing number of Web sites that list products containing RFIDs.

If you do purchase some product, such as a Gillette disposable razor, and it contains an RFID chip, then that chip belongs to you. You paid for it and you should be allowed to do anything you want with it, such as remove it from the product or, if that isn't feasible, destroy it.

Placing the actual chips in a microwave oven may not work; the chips probably do not contain the moisture (water) needed for the microwaves to resonate and cause heat. But if the product can be moistened without damaging it, then the oven might destroy the chip.

If the tag is in the packaging, burn it or include it in the stuff you recycle. But don't just toss it in the trash. OK, it may be stretching credibility to believe that the garbage trucks of tomorrow will have built-in sensors to keep records of the things you buy and throw away; likewise, it's unlikely that, as suggested in an article I once read, burglars might have portable scanners to analyze the contents of your garbage cans in order to see if you recently bought something valuable that they want and can fence. But anything is possible.

COLLECT AND TRADE, MIX AND MATCH

I read somewhere on the Internet about certain people who get together for regular meetings and later, when they go out for dinner and drinks, they take all their supermarket cards and toss them in a pile, mix them up and then everyone takes one at random. The idea is to confuse the computers that maintain a record of everything they buy.

Do the same with the chips. Collect them. Put them in some kind of little plastic container and carry it in your pocket wherever you go.

IF YOU WON'T JOIN 'EM, JAM 'EM

Sooner or later, if it hasn't happened already, someone will come out with a pocket-size device that confuses the reader. This will lead manufacturers to come up with a new model that uses frequency hopping (and then someone will defeat that), but regardless of who does what to fight these insidious gizmos, RFID is here to stay.

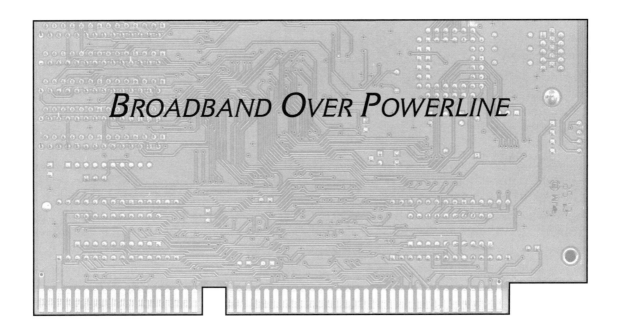

Broadband Over Powerline

Broadband over powerline (BPL) is a new system based on existing technology that will make high-speed Internet connections, as well as television and radio broadcasting, available through the power lines.

With such technology, it would no longer be necessary to run cable into homes and office buildings, and there would be much less awakening to a chorus of jackhammers tearing up the streets to lay fiber-optic cable. Another advantage is that broadband would be available in rural areas that are too far from the nearest central office (CO) to get DSL and where there isn't (yet?) wireless access.

While BPL has, as of November 2004, been set up in a few test areas such as Manassas, Virginia, it remains to be seen if it will spread to the rest of the country.

There are reasons why it may not, which incidentally will bring a sigh of relief to a great many people. For one, there are plans in the works to make wireless available anywhere in the nation; everyone will be within range of wireless access points.

And then there is the increased use of fiber-optic cable (FO).

Presently, FO does not necessarily enter individual buildings. It might go, in the case of telephone service, from a CO to major distribution points, but from there copper wire takes over. Telephone companies have been making "last mile" fiber noises for a long time that would change this, which would have the FO cable terminating within homes and offices.

And there are even plans to snake the FO cable through existing gas mains. Perhaps the proverbial meter reader would also determine your monthly Internet bill.

So FO and wireless may eliminate the "need" for BPL. Which would be most unfortunate for the feds, as they want BPL. Especially in metropolitan areas, where there are more people to spy upon.

YES, BIG BROTHER IS STILL LISTENING

Communications through this new medium, as well as FO and conventional copper twisted-pair wires, have to comply with CALEA, the Communications

Assistance for Law Enforcement Act of 1994. (It is at 47 USC § 1001-1021, should you want to look it up.) The law defines the existing statutory obligation of telecommunications carriers to assist law enforcement in executing electronic surveillance pursuant to court order or other lawful authorization.

And as we now know, the government has gone to great lengths to be able to read every single e-mail sent over the entire Internet and intercept every phone call using programs such as Echelon, Magic Lantern, and Carnivore. So, BPL can and will be used for the same purpose—to know everything we communicate to others.

HOW IT WORKS

The technology is similar to the once popular wireless intercoms that are now sometimes found in second-hand stores and thrift shops, in that the conversations are sent through the power lines. If you live in a large apartment building, you might intercept someone using one of these intercoms, especially if you buy the kind that have multiple channels. One of the types of surveillance devices we use on a technical surveillance counter measures (TSCM) sweep is called a "carrier current" detector that finds bugs that use this method.

Now, these intercoms work only with wall power outlets that are from the same power transformer, the ones that may be on a pole in the alley or underground and that step down 2300 or 7200 volts or whatever the distribution system uses, that comes into our homes. And it would be the same with BPL, so it is necessary to work around this limitation.

One way is to use a bridge that bypasses the transformer and sends the signal around it. Another is to install a wireless access point (AP) on the poles, with a modem set up in your home. Then, you could plug in a computer to any wall outlet and be online. No other cables, no other connections needed—plug and surf, as it were.

HACKING BPL

Now, as to the Internet, everything that connects has a unique number, just like telephones. This number is known as an IP (Internet Protocol) and is represented as a series of digits called dotted quad notation, such as 66.94.234.13. This is one of the IPs for Yahoo!

This could possibly include all of the power outlets in your home, since it is not unusual for a family to have more than one computer. The IPs could be set through the modem/router using DHCP (Dynamic Host Configuration Protocol).

As any hacker knows, an unsecured network can be easy to break into. And unless you, the BPL customer, know how to make your new home network secure, you will be wide open to hackers; what remains to be seen is how much access you will have to the modem/router the BPL provider installs.

If you buy an ordinary DSL router, such as the Speedstream that I use, then you have complete access to setting it up and to allowing access to your network only by those computers (your own) that you specify.

Will you have that option with the BPL equipment you are provided with?

I don't know; it is too early to tell. But it is something to consider when confronted with the choice between BPL and DSL or cable. There will be updates on this issue on the Web site for this book.

THE BPL PARTY LINE?

This is something else to consider. I will use cable TV as an analogy. The many different channels run through a single cable without interfering with each other because they are multiplexed.

One method is called frequency multiplexing, which means that each signal, or channel—HBO, History Channel, etc.—is sent through the cable at a different frequency. More on that coming up, but what this means is that it is possible for the company that manufactures the modem/router to build in circuits that would cause your Internet connection to operate on a second frequency—one that the feds listen in on; one that you aren't told about.

Another possible hazard is if (and I don't know this yet) different subscriber modems are multiplexed as with ordinary cable Internet. If so, then you are on a sort of party line and anyone with minimal hacking skills might be able to take control of your computer. Time will tell.

THE (BPL) COLONY?

There was a movie I rented a year or so ago, *The Colony,* about a subdivision where audio and video surveillance devices—cameras and microphones—were placed inside the private homes as they were built. Even the bedrooms.

BPL could become like that, and yes I am serious. The technology obviously exists, and anyone who doubts that the U.S. government will employ these devices is in need of a reality check.

The devices are small enough and being on the power lines, they require no batteries. And like the party line they could use a different frequency.

Also, since such devices will presumably need their own Internet IP, everything that is plugged in could be a potential "leak" to hackers.

Existing structures would not, of course, be pre-wired as in *The Colony,* so special wall outlets will be required to make the system work. And how would you know anything about this, other than what the installers tell you? Or the people who come around to make "maintenance tests"?

Audio bugs as small as a pencil eraser could easily be built into these wall outlets and light switches and, again, they would have constant power. The same is true of video and, of course, they could be built into, or later installed within, bathroom fans, clocks, radios, and anything else that plugs in.

RFI:
RADIO FREQUENCY INTERFERENCE

Another problem with BPL may be a disaster to ham radio operators who use high frequency (HF), the bands from 2 to 30 MHz or so that are capable of long-distance communications. BPL will use the same frequencies to send Internet data, and radio and TV signals as are used by amateur radio operators. But because these frequencies are being sent through ordinary wire, which is not shielded, those frequencies leak out and radiate into space for a certain distance. This can be very disruptive to such communications.

Now that may not seem like a big deal to most people, but during the Loma Prieta earthquake, I got involved in emergency communications, relaying messages with a handheld radio to operators who had this long-distance equipment. Many thousands of people were able to get in contact with friends and relatives across the country that otherwise would not have been able to.

And it isn't just us hams that use HF. The Highway Patrol in most states depends on HF, as their patrol cars cover a wider area and aren't always within range of a repeater, as are local police. And many private agencies such as the Red Cross, search and rescue teams, and local and state fire networks also use HF.

And if you like listening to short wave, to international broadcasting stations, and you are within a certain distance of the power lines that carry these interfering signals, then you may have to just give it up.

Current info may be found at the Amateur Radio Relay League, http://www.arrl.org, and there is a video file explaining this, with actual field tests, here:

http://216.167.96.120/BPL_Trial-web.mpg

You can view it with WinAmp.
So, what can you do?
If you don't need BPL, then don't use it. Although you cannot (yet) prevent the signals from being on the power line in your home and office, you don't have to connect any BPL devices. And if DSL is available to you, it might be a good idea to use it until more is known about BPL.

UNPLUGGING HOMEPLUG

Even without BPL, there are already a number of networking products available that work through the power lines, from manufacturers such as Belkin, Netgear, Linksys, and many others. The principles are, far as I know, the same and so are vulnerable to hacking. Just like wireless intercoms.

Check out http://www.homeplug.org/ and Just Say No.

Again, BPL is fairly new, and so far has been implemented in only a few areas. No one knows yet how well it will work out. And there are many unanswered questions. By the time this book is published, much may have changed. So you are invited to visit the Web site for this book at http://www.fusionsites.com/dbm2 to get updates on what might be the most insidious and pervasive method of spying on We The People that the government has ever implemented.

# Part Two

ELECTRONIC SURVEILLANCE

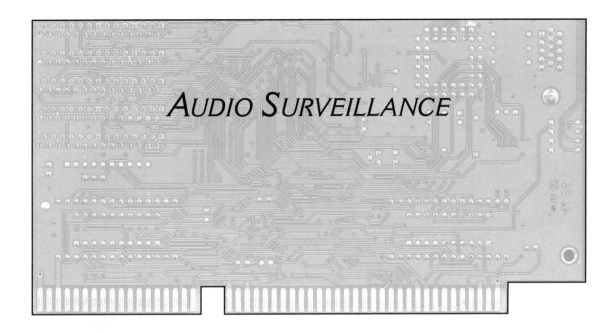

Audio Surveillance

Audio surveillance has been covered in my previous books and some of coauthor Steve Uhrig's articles, so here I will go over just the basics. You might also read David Pollock's book listed in Appendix B.

Audio was, obviously, the first method of eavesdropping known to mankind. It evolved from listening under eaves, to placing a water glass against a wall as a makeshift stethoscope, to the modern methods in the electronic age.

One such modern device is the directional microphone, such as the types used by sportscasters at football games. They are very expensive (the good ones run a grand or more) but are available to anyone who has the cash. Sennheiser is supposed to be the best, and Shure or Electro-Voice are also excellent; however, they are subject to wind noise and generally overrated as to distance.

There is another type of highly directional microphone that a person handy with tools can make himself. I have never built one—I've always wanted to but never seem to have the time. It consists of a series of tubes made from aluminum or stainless steel (or possibly plastic) about 3/8-inch outside diameter cut in 48 or so pieces from 1 inch to 48 inches in length. Looks a little like a Gatling gun. All of the tubes are flush at one end, bound together with straps or bands, and a housing of some kind—a funnel or something like that—is fitted over the flush ends. The openings are sealed with silicon or liquid rubber, and a microphone is placed inside the funnel.

The tubes would have to be sturdy enough to prevent them from bending, and they would have to be perfectly straight to begin with. I would like to hear from anyone who has ever built one of these things.

But this device, too, is overrated and subject to wind noise.

Parabolic reflector types do work but to be really effective at any distance, they need to be the size of a satellite TV dish—6 feet or so in diameter. This is because of the low frequency and long wavelength of audio. It's not practical for covert operation, the cost is much more than a grand, and they, too, are subject to interference from wind.

Unless such a device is behind acoustically transparent material or otherwise hidden, if it can hear you, then you can see it.

CELL PHONES

Technically this should be considered RF surveillance, a bug. Some models can be configured to automatically dial the number of a second phone, which will automatically answer. But there is no indication that this is happening. No clicks, beeps, or flashing lights—the LCD display is blank, as if the phone was not even turned on. So if someone leaves a phone behind after an interview or any important meeting, there is always the chance that it might be a short-term bug—till the battery runs down.

A well-equipped operative knows this. Did you?

DIGITAL RECORDERS

Smaller is better when it comes to using audio recording devices in most situations, but especially in surveillance. Once upon a time, someone managed to get a bulky wire recorder up on a telephone pole. Before magnetic tape, recorders used steel wire about the size of fine fishing line, and the device itself was about the size of a breadbox. How they did it isn't known except to them.

Meanwhile, digital recorders have gotten smaller and less expensive, and now there is the DR-7 available from Shomer-Tec (http://www.shomer-tec.com). It is about the size of a pack of gum, can store as much as nine hours of audio, can be placed in voice-operated mode, naturally, and costs only $109. And it can even dump the audio to a computer for storage, or to send as an e-mail attachment.

LASERS

Q: Is it true that lasers can be used to hear people talking through closed windows?

A: Yes, but....

We built one of these systems in college lab. Here is how it (sometimes) works.

A low-power laser is focused on the target window. The return (reflected) beam is received through a small telescope and focused onto a circuit that is connected to an audio amplifier. The sound in the target room causes the glass to vibrate, which makes a tiny change in the distance between the glass and the laser system, which is converted back to audio.

But precision is everything with this system. When a beam of light strikes a surface, it is reflected at the same angle at which it strikes. (Shine a flashlight straight ahead into a mirror and the beam reflects back at you. Stand against a wall and hold it at a 45-degree angle and you'll see the beam reflected on the opposite wall.) So, obviously the laser has to hit the wall dead on to be reflected back to its location to be captured by the receiving device. Otherwise, two locations would be required, which is not easy to set up and would also impair its performance. Also, they work only under optimal conditions. Fog, snow, rain, passing cars, aircraft, elevators in the building, all interfere with the return signal.

Lasers are easily defeated. Close the drapes. Place a fan near the window. Tape a small transistor radio to the glass and treat the spies to a little Beethoven. If you want to see for yourself, you can buy one from Information Unlimited in Amherst, New Jersey. They have some really interesting stuff there, but don't expect it to work as well as in the movies.

THE INFINITY TRANSMITTER

The original infinity transmitter was actually a room audio surveillance device, meaning that it sent the sounds made in the room where it was installed through the phone line. It used a hook switch bypass, which shunted across the hook or cradle switch, turned the microphone on, and prevented the phone from ringing. To activate it the user called the target number and, before it could ring, fed an audio tone into his phone. Since the Telco ESS system does not connect the two phones together until the called party answers, the old infinity transmitter will not work like this.

Another type of infinity device was once made by Viking International in San Francisco for use on a fax line. The phone may still ring, or beep, depending on the type of fax machine, and anyone in the vicinity of the machine will be alerted that a "fax" is coming in. But when it doesn't hear the handshaking tones, the fax gives up and after about 50 seconds it times out. Normally, this would free the line, but the Viking device keeps the line open and the microphone turned on. I borrowed one from the owner and installed it on my fax line and called it from the voice line to see. It worked fine after I remembered to adjust the gain control. Viking's Web site is http://www.vikingint.com/.

Like most other surveillance devices, they are disappearing in the United States, but there are still some available if you know where to go, and of course there are foreign markets but be aware that Customs will probably confiscate them.

As you will read later, any inside audio listening device can be found in the physical search, if the search is done right. Anything that could conceal a microphone—and this can be virtually anything—can be examined. And unless it is attached to a transmitter, it will have wires connecting it to the listening post or a recorder that can be accessed periodically to obtain the information and reset or install a new tape.

Learn to know what kind of wires should be in your home and what should not. Keep in mind that they can be very small, even concealed under a coat of paint.

Next, if there is a microphone, what might the person who is listening hear? The answers might surprise you!

WHAT A SPY MIGHT HEAR

In *Don't Bug Me* I described how surveillance works in the movies. The good guy/bad guy installs the bug and instantly has crystal clear audio—from an improbable distance—and naturally hears exactly what information is needed. In real life, of course, it doesn't work that way.

Once the equipment has been installed, a team of spies working in shifts live a life of mostly boredom at the listening post, waiting for days or months and maybe still never get the evidence they need. At least not in so many words—unlike our bad guy in *Don't Bug Me*, the real-life suspect may never come right out and say, "OK, youse guyz, we rob da Foist National Bank at 9 A.M. tomorrow." Or whatever they are supposed to be up to.

An experienced operative learns to listen to all of the sounds that come through the headphones—not just what people are saying—to analyze them and see what they might add up to. First, as an example, let's consider the nonvocal sounds made in your own home on several different days.

Suppose you were to sit in your favorite living room chair, blindfolded, from early morning till midnight on a weekday and a Saturday. And suppose the members of your family, who you managed to get to go along with this nonsense by bribing them with offers they couldn't refuse, spend both days without speaking. Not a peep. It would no doubt surprise you how much you were able to learn of what is happening in your own home. Like a sightless person, you would come to depend more on hearing, to develop it.

Now, let's look at some of what goes on during a typical weekday and a Saturday.

You hear an alarm clock going off, and from the direction you know it is one of the upstairs bedrooms. You hear someone trotting down the stairs, the fridge being opened and closed, then the front door slamming. OK, it was the first of several alarms, and whoever got up didn't turn the shower on or spend much time in the kitchen, then out the door. Logical conclusion: It is your oldest boy, at 5 A.M. getting ready for his paper route. And although you hadn't thought about it at the time, you recognized the sound of his footsteps. Trotting, rather than walking.

Someone else gets up, and at this point, you pretty much know who it is based on the order in which people get up. But that's not good enough to draw a conclusion—you're not an operative quite yet. In the real life of an operative, you wouldn't know the bad guy family routine. The shower goes on, followed 15 minutes later by a hair dryer. Feet pad down the stairs, you hear the clink of silverware, and a few minutes later a teakettle whistles.

Who uses a blow dryer and drinks tea? Your daughter, getting ready for school.

Mr. Coffee starts to gurgle and you know it is the time you and your wife usually get up. You hear her come down the stairs, the footsteps faint, as she is wearing slippers.

Your son gets back from his paper route—you hear the screech of his bicycle brakes, the door opens and slams shut. He gets ready for school and then out the door.

Later, the TV comes on. *Days of Our Lives* or whatever—you suffer through it, the kids get home from school, feet running up and down the stairs, a computer game, the clatter of dinner being prepared, the chiming of the microwave, and the sound of evening TV shows.

Saturday.

The routine is different. Instead of your wife being awakened by the clock radio, it is the blasting of cartoon characters on the television.

A lawn mower starts up next door, punctuated with the ringing of bells from an ice cream truck.

And then, what you don't hear.

Tommy must have found someone to deliver his papers today; otherwise, he'd have been gone. The teakettle doesn't whistle.

So, think about how much you would know on a given day if you couldn't see anyone and no one was speaking.

Now, let's look at a surveillance situation. You and one other operative are sitting in a small rented room across the street from where a suspected bank robber is holed up, perhaps making plans for the next heist. You have his NCIC and local police department files, which you have studied carefully to learn as much as you can about the subject: his habits, preferences, and eccentricities. And it is this information you will use to determine if it is the right person as well as details of the next crime.

You know that the suspect usually works alone. And that he is too smart to use a phone, especially a wireless type, so he isn't likely to speak about the crime out loud unless he talks in his sleep.

The apartment the suspect is in is at the rear of the building and the only windows face that way, to the back, but the entrance is on the other end, the front.

You have been on the stakeout for a week, and while you have other operatives driving past the front entrance, you cannot use a surveillance van on the street, as the subject is likely to make it if he comes and goes.

So, as far as you know, the suspect seldom leaves the apartment.

Reading the files, you take a lined pad and write down a few things to keep in mind. To listen for. What sounds you should hear, if it is the right person, and what sounds are conspicuous in their absence.

The guy is a two-time loser and is known for carrying as well as using weapons. So, assume that he is armed, and with something other than a zucchini. So he will more than likely check his weapon(s) periodically and you may hear the unmistakable sound of a shotgun slide being racked or a magazine being slapped into an automatic. But if you hear this shortly after the grunt of pulling on a pair of boots or just before the sound of the door slamming, you conclude that he has left with the weapons. Time to get ready, call for backup.

If the bad guy is known to use radios to monitor the monitors, you may hear your own guys on his radio. But what if he is using earphones? You may not hear the radio. On the other hand, if his batteries get low, you may hear the faint beep, beep, battery warning indicator. This is a sound you need to learn to listen for. Is there a commercial radio playing? Do the files indicate a preference in music? A particular type of music? Do you hear it, or something different?

Do you hear a knock on his door periodically? The door opening and closing? A few muffled words such as paying a pizza delivery person?

So, while in real life you will hear people speaking, what they have to say may not reveal as much as what they do not say as well as the background sounds that you have been trained to listen for.

Listen…

And then put it all together.

Sounds…

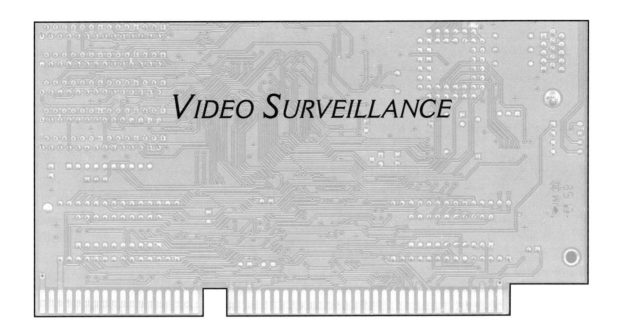

Video Surveillance

Unless you live in a vacuum, you are aware that the use of video cameras is exploding. They are *everywhere*—on the front entrances to office and government buildings and, increasingly, at apartment buildings and private homes. Businesses use them to monitor their employees, and they are starting to become standard equipment in elevators.

They're on the street in most large cities, on the corners of buildings or inside fire alarms, or disguised as power transformers on utility poles in the alley behind your home. They're in the workplace, the shop, the office, and even in the restrooms.

Video cameras have become so small that they can be hidden virtually anywhere, including inside baseball caps and even built into the frames of eyeglasses. And it had to happen—they are built into some cell phones as well as wireless PDAs.

Damn near anywhere you go you are likely to be within view of one or more of these insidious devices. Yes, I said in *Don't Bug Me* that there is no reason why a person should be under video surveillance in his or her own home. That was then. This is now.

They're small, they can see in near total darkness, and they're cheap. Used to be that you couldn't log on to a Web site without getting hit with one of those ubiquitous banner ads for the "X-10."

I believe that there are laws that restrict the use of video cameras in some workplaces, and your employer is not, apparently, allowed to install cameras in restrooms, locker rooms, etc. But then, people aren't allowed to rob banks...

So what can you do? Well, not much other than being careful what part of your body you scratch when you are in an elevator. Hide. Get a work-at-home job. Wear a disguise whenever you go to the market?

Seriously, there is little you can do to avoid them, although now and then on the Internet you can find Web sites that have mapped their locations.

Disabling them with a laser is flat-out illegal and can get you tossed in jail.

What can you do? Outside, learn to live with it. At work, learn to live with it.

But in your own home, remember that the physical search will—if done right—

uncover literally any surveillance device. A laser pointer will disable them, so if you use one make sure you cover every square inch of walls, ceilings, and even floors from many different angles.

WIRELESS VIDEO CAMERAS

Don't be the only person in your neighborhood without the X-10, the Internet ads used to say. Indeed, many thousands of these tiny cameras were sold before the company went out of business. And some of these cameras may be watching you!

The obvious reason for using wireless is that because they need no wires, they are quick and easy to install. Just plug the transmitter in, select the channel, and point the antenna in the direction of the receiver. And, as at least some of them use spread spectrum, the signal is less easily blocked than ordinary types of transmission and will not be detected using conventional countermeasures equipment.

Now, I have never had much interest in video, but then someone sent me an article about how a guy in San Francisco was using one of those little portable TVs and a homemade antenna and walking around to see how many of them he could pick up.

Hmmm, fascinating. I wanted to try it. I didn't have an X-10 so I found one in a pawnshop for $25. But I didn't have any kind of small TV to walk around with. So, I started thinking that since there are probably a lot of these wireless cameras, and since I do have the X-10 and a big 2.4 GHz antenna, maybe I can pick up the transmissions of a few of them here in my neighborhood. The natural born spy in me takes over and I envision intercepting all sorts of juicy stuff. Sordid bedroom scenes, maybe someone murdering someone, or at least a bank robber. I will be famous: "Alert citizen foils bank holdup." The mayor will thank me on TV. A good citizen award. Offers of jobs, of marriage.

The receiver has an ordinary phono (RCA) jack, but my antenna uses an "N" connector, so I pulled the X-10 apart (which, incidentally, was a bitch as the screws wouldn't come loose so I ended up chewing the plastic case apart till I got it open), then I removed the cable connector from my high-gain antenna and soldered it to the place where I removed the small built-in antenna the X-10 uses. The receiver has a composite video output (also RCA or phono) jack, and I had a patch cable that would connect it to the video input of the main desktop computer.

At this point I ran into a problem. I had somehow lost the setup disk for the video card, a Rage Fury Pro, and so I had to download the MultiMedia center programs from the ATI site. This turned into a real hassle, as the names of their video cards are confusing to begin with (where the hell do they come up with the strange names they use?) and they apparently aren't much into tech support. Of all the e-mails I sent, the only answers I got were from autoresponders—never a personal reply. And trying to get them on the phone? Either call their 900 number (very expensive and you get put on hold) or pay something like $40 per call paid by credit card. Now, ATI makes some excellent products but if you buy one, whatever you do, don't lose your setup CD.

Once I finally found and downloaded the Multimedia Center and got it installed, and connected the receiver to the X-10, I was ready to spy on the entire neighborhood. Hot damn! A quick trip to the fridge and I pop open a bottle of Heineken and fire everything up and start scanning.

AUDIO SURVEILLANCE WITHOUT A WARRANT is still technically illegal. But if federal agents install a video camera in such a way that they can see people's faces, they could employ skilled lip readers to extract at least part of what is being said. Little doubt that they have been doing so for years. It may or may not be admissible as evidence, but this is one of those situations where they use unlawful surveillance on someone they suspect of something, and when they obtain sufficient information, then they apply for a surveillance warrant. These warrants, incidentally, are rubber-stamped. According to *Wiretap Report*, which is published every year by the Administrative Office of the United States Courts in Washington, D.C., not a single application for a court order has been denied by a judge in several years.

Close-up of the X-10 after I opened it to connect the antenna.

Another view of the X-10, connected to the big grid antenna.

This was a slow and somewhat tedious process, carefully moving the antenna from left to right, then changing the vertical angle and repeating, for each of the channels the receiver works on. The grid antenna is highly directional. So I was very careful ... and I saw ... nothing. Not a single bedroom, active or otherwise, to be found. Nor did I see anyone committing a murder. Not even a dreary deserted parking lot. Literally, nothing.

AHA! I said to myself, I will try it with the antenna pointing out the west window. Maybe I will get to see "college girls take it all off."

Nope. Nothing. I was truly disappointed. While I didn't really expect to catch a bank robber, I did think there would be at least some shadowy images of something or other. Nope.

One of the reasons I didn't get any images could be that the antenna for the transmitter—assuming there ever are any in this neighborhood—is pointing to the receiver, which is inside the area under surveillance. The camera is focused on a parking lot but the antenna sends the signal to the booth where the attendant works during the day. Or a camera that monitors the lobby of an apartment building has the antenna pointed to the manager's office and not toward the outside world.

Well, I had considered that but figured that with the high-gain antenna I was using that I still might pick up some images.

Nope.

Now, the subject of this article (the source of which I have misplaced) had some success in getting images on his portable television because he was wandering around the downtown area and Financial District. But I was in a part of San Francisco that consisted of apartment buildings and small independent businesses.

So if you use one of these cameras to protect your home, it isn't as likely as the article implies for snoopers to capture what it is capturing. Just remember to have the antenna pointed away from the outside, away from the street.

A company in the UK called VideoScanner (http://www.videoscanner.co.uk/) sells the monitor pictured on their site. It scans through the entire 2.4 GHz band, looking for wireless video transmitters. And GLMFG (http://www.glmfg.com/website/microwave_video_rx.html) offers another video receiver as well as transmitters, amplifiers, and antennas.

RADIO FREQUENCY SURVEILLANCE: BUGS

Who knows of the first use of an audio radio transmitter by one person to spy on another—when it was or why? But it is probable, given human nature, that it happened shortly after Lee DeForest invented the triode, the first amplifying vacuum tube. It might have been the size of a wire recorder breadbox and hidden in a kitchen cabinet with a bucket full of batteries, but whatever the details, a new era in spying was born.

The Bug Book, one of my previous books with Paladin Press, is entirely about radio frequency surveillance. This 157-page book has much of what there is to know about the subject, at least as of when it was published in 2000, so I will not rehash all that information here. As I write this four years later, there have been some new developments, but essentially a bug is a bug and they all have some things in common:

- They transmit a signal.
- That signal contains intelligence—the sounds of people's voices.
- It has as much power as needed for its signal to reach a place where someone is monitoring—the listening post.

Other considerations are that it is hidden well enough to be difficult to find and that it has constant power so there is no need to enter the bugged area to replace the batteries.

Bugs are subject to the laws of electronics and physics and wave propagation, and while this varies with the type of modulation somewhat, still a bug is a bug, and new developments and design in surveillance transmitters still have these things in common. So if you want to learn about surveillance transmitters, then *The Bug Book* is still all you need.

Meanwhile, a few comments and then a few words on some new developments. Surveillance, RF or otherwise, in real life is rarely as depicted in the movies or TV—bugs as small as a sugar cube that pick up sound from 100 yards away, can transmit several miles, and operate six months on a calculator battery? Uh-uh, no way.

Under perfect conditions (which are extremely rare in the real world of surveillance) a surveillance transmitter could have a range of miles, but it would be considerably larger than a sugar cube. A typical transmitter will be 1 to 2 inches square and half an inch thick. Then consider the batteries. A calculator battery might power a very small transmitter with a probable range of 100 feet or so for a few hours, maybe a day. The VT-75 transmitter from Deco Industries, as I recall, has an output power of 7.5 milliwatts and runs for about two days on a 9-volt battery. Again, read *The Bug Book*.

WHAT IS NEW

Tech stuff: single photon emission.

A photon is a particle of light or a particle that sometimes acts like a wave; depends on which physicist you ask. And if it were possible to build a device that generated and emitted single photons, these particles could be used to transmit information. Once this was only a theory, but it is becoming a reality. The following is from the article "UCSB Professors Build Communicative Device With New Photon Applications" by Josh Braun, which ran in the *Daily Nexus* (the UC Santa Barbara student newspaper) in January 2001:

> "Single photon emission also makes possible a method of encoding secret messages, known as quantum cryptography. This system is based on the idea that photons are the most basic unit of light and cannot be divided. ... It is impossible to observe a photon individually, because to see something requires bouncing a photon off of it, and bouncing a photon off another photon makes them both change course."

Think of it like this: You have a spy transmitter that sends its signal on this beam of single photons, sort of like modulating a light beam except that it is so concentrated that only single particles of light (sort of like a laser) are transmitted.

As long as the beam makes it from its origin to its destination, fine, the information it carries gets there. But if someone places something within the beam, it is disrupted; it stops working. Therefore, while it might be possible to intercept—to eavesdrop upon—the signal, by detecting it you have interrupted it. Stopped it. Sort of like using a mirror to reflect the light away from where it was originally shining. On the other end, the listening post, the operative there knows this, of course, and so shuts down the operation.

So what this means is that while it is possible to defeat such a system, and while the equipment can be found during the physical search, it is nevertheless not possible to intercept the information being transmitted.

You can't spy on the spy.

SPREAD SPECTRUM

Actually, spread spectrum (SS) technology is not new. It was developed back in 1941 to prevent the enemy from jamming the radio signals that controlled Allied torpedoes. The technology is credited to actress Hedy Lamarr, who first envisioned it, and composer George Antheils, who developed it into a workable system.

More and more communications devices are using SS, including cordless phones, wireless networks, and, naturally, surveillance devices—bugs. They have greater penetration than some AM or FM devices and are difficult to monitor.

So how does spread spectrum differ from other types of radio transmission? In two ways; there are two types of SS transmission. The first is frequency hopping, in which the transmission, the information being sent, is broken into segments that move from one frequency to another in a pre-arranged scheme. If, for example, there were a TV program that had the first five minutes on one channel, the next on another channel, and on and on, you wouldn't be able to watch it unless you knew which channel to go to next and were able to switch that fast.

The other is direct sequence, where a single frequency is used. The information—such as in wireless networking—is sent in segments or packets in a sort of redundant error checking method.

This is, of course, an oversimplification but I think you get the idea.

Spread spectrum signals don't show up on most countermeasures equipment, the spectrum analyzer being one exception. But still, the operator has to know what he is looking for. And with microminiaturization and custom-made chips, SS is no doubt implemented into body wires and, naturally, bugs. So, where we ordinary humans had at least a sporting chance to find bugs that used frequency modulation, with SS the only way we will is through the physical search or using thermal imaging, described in a coming chapter. I don't know of any commercial sources of SS surveillance transmitters. There may be some in European countries where the laws aren't as strict as in the United States, but importing them is very illegal.

A good electronics technician could probably modify a cordless phone to work as a spy transmitter, and there are chipsets available from many sources for those who have the ability to build their own.

So, as I have stated elsewhere, if someone seems to always be one step ahead of you, to know things he is not supposed to know, or if a competing company always manages to underbid you by the smallest of margins, then maybe you are the victim of a bug.

Again: The physical search will unearth virtually any inside listening device if done right, and you can learn a great deal about how-to in *The Bug Book*. Now, on to telephone systems, wiretapping, party lines, and a little humor about the early days of Mr. Bell's invention.

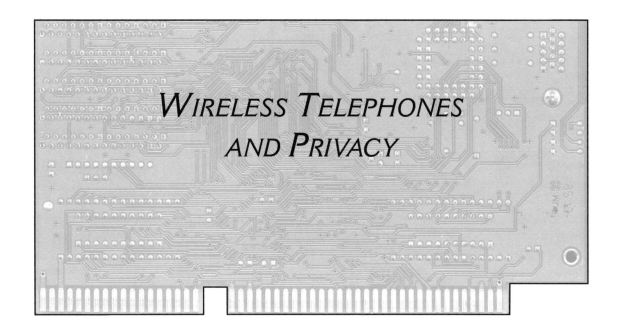

Wireless Telephones and Privacy

I REMEMBER A CONVERSATION I HAD WITH AN ATTORNEY SOME YEARS AGO. He had hired me to unpack and set up a new computer, and we got to talking about one of the books I had written. I don't recall just how the subject came up, but it came around to telephones and so naturally I started ranting about phones without wires.

He didn't realize that a cell phone is a small two-way radio station. Or that some people could "tune in" and listen to him, hear his conversations, record them, maybe use the information to their advantage. Many of the analog cordless phones that people can't seem to live without have this drawback too.

In *The Phone Book* (also available from Paladin Press), you can read about wireless phones in great detail, so I will not repeat all that is in that book other than to go over the basics. Yes, indeed, wireless phones are two-way radio stations. If you are using a standard analog cordless phone, anyone with a scanner or communications receiver that tunes the cordless channels might well overhear you. Virtually every scanner made covers the 46 and 49 MHz frequencies they use.

The new 900 MHz (902–928) analog phones are no more secure than the VHF models; they just use different frequencies.

However, some models use digital signals, and others, spread spectrum. Digital cordless, like digital cellular, will defeat the average listener—the "kid with a scanner" and virtually anyone else except for government agencies who have the technology to convert digital back to analog. Yes, it is true that a determined person—a private person or agency—might be able to build a complicated device to convert the digital transmission back to ordinary analog speech, but this isn't a trivial matter.

Spread spectrum, as you read above, can be very difficult to decode depending upon the type. It takes sophisticated equipment, special software, a very fast computer, and a great deal of determination. But keep in mind that the NSA is not going to allow us to have any form of communication that is so secure that they cannot eavesdrop upon us.

But keep in mind that while the NSA does not want us to have any form of communication that is so secure that they cannot eavesdrop upon us, they still do not have the manpower to tune in on every call from every phone. Even with automated systems such as Echelon.

Under some conditions, however, this is possible with the old analog system, which some people still use. It depends on the location of the target phone, location (and type) of the receiver, antenna, and the skill of the operator. In researching the first edition of *The Phone Book*, I was able to zero in, several times, on a particular phone that was located nearby (about one block away) and also to track conversations. One, I followed for several hours through a dozen changes in location. It was a man in a Jaguar calling different 900 "sex line" numbers. Must have cost him a bundle! See *The Phone Book* for details.

But, you say, when I bought my phone I was told that no one could listen in on me. Was I lied to? How could this happen?

In the early days of the cellular system, the mid-1980s, the person who sold you the phone didn't necessarily lie to you; he may have believed, have been told, that your calls are private. He may not have known the truth about the cellular system.

If you were told this in the last few years, then, yes, they lied to you. The vendors know well that analog cellular can be easily monitored. But—would they have told the truth if it meant losing a sale?

As to how this could happen, it is very simple: The people who developed the cellular radio system were interested in making money, not protecting your privacy. They did not learn from the mistakes, the weaknesses, of the early mobile telephone service. If they did care, they would have made the system digital in the first place, something that they are doing now with the PCS, personal communications system.

If people, "kids with scanners" or whomever, can monitor my cell phone conversations, then they might overhear my credit card or calling card numbers, right?

Yup.

Well, then, they can also overhear my voice mail password and get into my mailbox, right?

No. If the people monitoring you have a DTMF (touch-tone) decoder, they may get your password, but that is of little use to them as they don't know the number you call to access your voice mail. I will explain:

When you make a call from an analog cell phone, the number you "dial" goes out over the reverse data channel in a code called Manchester. An ordinary scanner or communications receiver does not decode this; anyone listening on the reverse data channel hears only a buzzing sound. A special device is required to convert this Manchester code into plain English letters and digits.

Such devices, as described below and in *The Phone Book*, exist, and before the feds cracked down, thousands of them were sold.

Will cellular phones ever be 100 percent secure against eavesdropping?

No. Not 100 percent. Nor will any mode of communication for that matter, excepting the single photon technology.

Using a digital phone will defeat scanners, communications receivers, and most intercept systems, but it won't be long before someone starts marketing an inexpensive system that will convert the digital signal back to audio. Such a device is not easy to build but there is a market for it, as so many people are using digital rather than analog cell phones. Sooner or later it will be available on the black market.

Digital transmissions can be encrypted to make them more secure, even unbreakable, but far as I know, none of the cellular vendors offer this. It may be that the cellular industry will not be making any major changes due to the personal communications system that is currently being developed. If PCS does offer digital encryption, it will be secure enough that the only person or persons who will be able to break it is someone with enough money to build a computer specifically for that purpose, such as the government, a large corporation, or possibly a determined technician with a lot of time and money.

The code division multiple access (CDMA) system that is in use in some areas is the next step up in secure telecommunications. It is very difficult to break, requiring an expert with access to a very fast computer—a mainframe would probably be required, or perhaps a Sparcs workstation—as well as a great deal of time.

This is one reason the FBI and other agencies are pushing Congress so hard for laws to let them expand their wiretapping abilities; they want to be able to intercept any cell or PCS phone at the switch where the conversation is not "scrambled."

Another approach to monitoring "secure" cellular telephone is in the phone itself, rather than radios and commercial monitoring systems. No matter what the cellular companies use to make your conversations secure—whether it is spread spectrum, or digital, or encryption, or any combination of these—the fact remains that in order for you to be able to hear the person you are talking to, your phone has to be able to convert all this scrambling and whatever else back into clear speech. The phone you are using has to have that capability. And whatever is in your phone, whatever circuits are used, whatever technology is present, it can be duplicated.

Now this is no trivial matter. It will take a great deal of reverse engineering to copy this technology, but it can be done and it will be done. It is true that the chips in your phone can be altered—a "link" inside the chip can be burned open or destroyed, so that it is impossible to copy what is in that chip. However, it is still possible to use a sophisticated logic analyzer to record the signals going into and coming out of those chips. And from there, it is possible to duplicate what is in that chip, and then make copies, or clones, of that chip.

It is true that what the phone does to convert the signal back to plain speech may depend upon certain information that is sent to the phone by the cellular system computer. There may be some sort of code that changes with each use, such as shared secret data. In such a case, it will be necessary to intercept the signal, analyze it, and determine what is being done. Another nontrivial matter, but it can and will be done eventually.

The only way to have absolute security from eavesdroppers is to use an encryption algorithm that cannot be broken by anyone, such as Des-X or PGP. But since even agencies such as the NSA cannot break these, the federal government is working hard to prevent them from being used except by themselves.

Defeating digital and spread spectrum technologies are, again, big-time projects. But there are, or will be, a lot of people working on it. Independent hackers all over the world will figure out bits and pieces. Parts of the puzzle will leak out from the cellular vendors, phone and chip manufacturers, technicians, subcontractors, etc. Tons of useful information can be found in Dumpsters and overheard in bars and restaurants. And over a period of time, all of this information will be passed around and available to whoever wants it.

Cellular telephone calls will become more and more difficult to eavesdrop on, as will probably the PCS, but they will never be totally secure.

You really want total security? Try a rowboat in the middle of a large lake, equipped with sonar that detects scuba divers and remote-controlled submarines. Or Maxwell Smart's Cone of Silence. Or better yet, stop using telephones of any kind. Use electronic mail and encrypt it with PGP.

The commercial monitoring systems mentioned here range in price from about $500 to $35,000. Some are designed specifically for cellular monitoring (and advertised as such) and are unlawful to possess. Others may qualify as "test equipment" and therefore may be legal to possess. This falls in a gray area of title 18 USC section 2512, and a case in process in U.S. District Court in the San Francisco area may determine what is and isn't legal to possess and sell. The act of monitoring is unlawful.

The company where I was once employed as a technician (Tech Support Systems/CSI) manufactured cellular monitoring or "test" equipment. I built, tested, and demonstrated several types, which is why I am familiar with them. One model, called Cellmate, uses a modified Panasonic cellular phone. The target number is programmed into the modified memory location and when that phone makes or receives a call, the system locks onto it. It can also be set to automatically activate a tape recorder. However, the target phone has to be within a certain distance, and there is usually a one in three chance of making the interception. Other types are capable of scanning all (three or four) of the control channels in a given cell so the hit rate is close to 100 percent. They may be built from scratch (a fairly complex operation) or from modified cellular phones interfaced to a computer. For details, see *The Phone Book*.

If you still have an old analog phone, replace it. This system, digital, is somewhat more secure against both eavesdropping and cloning, but it, too, can and has been defeated. (More on this the next time I update the FAQ.)

And remember what you read in the chapter on privacy: The new cell systems use "microcell" technology, meaning that there are more cell sites, closer together, and using less power. So this makes it easier to determine the physical location of anyone using such a phone. Like a cell phone, as long as it is turned on, it is constantly communicating with the nearest cell site. A record is kept of this (for "billing purposes" only, of course). Actually, these records are available to law enforcement with a subpoena, but eventually they will be accessible simply upon request. So everywhere you go, you leave an electron trail, a record of your footsteps. If you don't like this idea very much, then don't use PCS. Or cellular. Do you really need either one? If so, turn them off when they are not being used. Have incoming calls directed to your pager.

Or even use an old-fashioned answering service and return messages from a pay phone. On the Web site for this book, you can read about something called Bluetooth that is a serious threat to cell phone security.

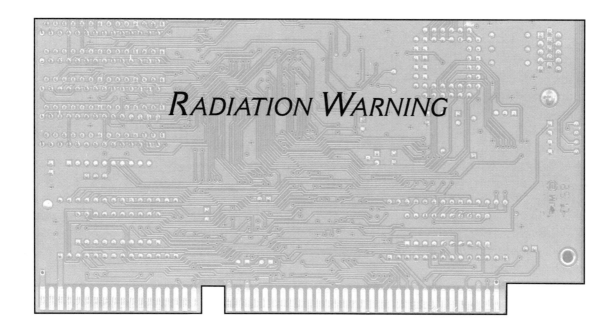

Radiation Warning

It has been known for decades that radio frequency radiation affects human tissue. To what extent depends on the frequency, intensity, and amount of exposure. Essentially what it does is to produce heating, although it is more complicated than that.

Now, I suspect that most people think of radiation as radioactive fallout from nuclear weapons and do not understand that radiation (along with radio and television signals) is part of what is known as the electromagnetic spectrum. It extends from a few thousand cycles (Hz) per second, which is audio, through the radio frequencies—AM, FM, TV—and to infrared and ultraviolet, then to X-rays, and finally gamma rays, which are produced by nuclear reactions. They are all part of the spectrum. The higher the frequency, the more effect the waves have on people.

Consider microwave ovens. They will heat virtually anything that contains moisture, which is why they will cook a pound of ground sirloin but have little effect on Rye-Krisp.

Microwave ovens and wireless computers operate in the 2.4 GHz band. At this wavelength, it is called "ionizing" radiation, which means that it is capable of causing molecular changes in human tissue. And, so I am told, not just heating—it can actually cause changes in the DNA of a cell.

Now, the amount of radiation from a 30-milliwatt card such as the Orinoco Gold classic with the built-in antenna is so tiny, and you are so far away from it—typically at least 12 inches—that your exposure is probably little more than the background (ambient) radiation. And even over a long period of time, I suspect that the danger is near zero. The power of radio waves decreases by inverse square, so the farther you are from the source, the less the intensity.

On the other hand, a high-power card with a directional antenna …

A few months ago, I sold a Senao card along with an 18-dB flat panel antenna to a guy at one of the geek meetings we have here in San Francisco. He had it running on a notebook computer and after a while experimenting, was talking to the others in the group while holding the antenna against his chest.

I noticed and suggested that he not do this. Even though the power is only 200 milliwatts, with the directional antenna it is focused, or concentrated. He mentioned that he did feel a slight warming. Now consider the 24-dB grid antenna I use, pictured elsewhere in this book, along with the 200-milliwatt Senao card. No way am I gonna stand in front of it.

Yes, high-power cards and directional antennas may produce radiation that is harmful. To what extent this is true is something to be determined by someone with more knowledge than I have, but common sense should tell you to keep the antenna pointed away from yourself or any other person.

The same may be true of cell phones, as explained elsewhere in this book.

Also, please keep in mind that if you will be using directional antennas, there is the possibility of interfering with other networks. If you happen to be near a hospital, for example, where wireless networks are used for monitoring the vital signs of patients in intensive care units, you could possibly interfere.

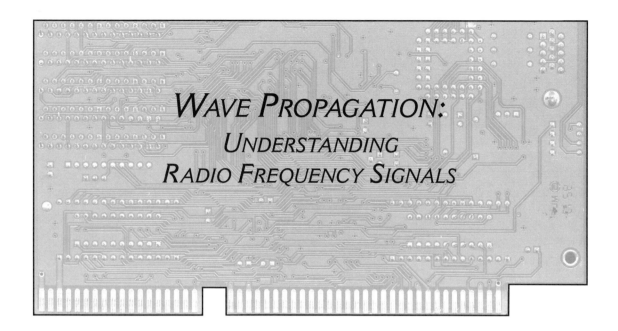

Wave Propagation: Understanding Radio Frequency Signals

THE CHARACTERISTICS OF RADIO WAVES—THE TRANSMISSIONS, THE SIGNALS from any kind of transmitter (AM, FM, and wireless fidelity, better known as Wi-Fi), what they do, how they act—are known as wave propagation; it is a very complex science about which many books have been written. Radio waves act differently and unpredictably depending on factors such as frequency and type of modulation, and they are subject to weather, sun-spot cycles, and even the time of day, as they tend to be "beaten into the ground" by the sun.

What this means to you who are learning about wireless computing is that the signal strength and speed of data transfer to the access points (AP) you monitor or attempt to associate with (connect to) may vary, fluctuate from one minute to the next. The AP can determine the strength of the connection; if it is weak, data will be sent at a slower rate.

Another important factor is that anything in the signal path can cause multipath distortion (ghosts). If you have a television set that is not on cable, that uses rabbit ears, then you may see that there is a second image on the screen. Let's say that you are watching the talking heads on the 10 P.M. news. One of the newscasters has what looks like a shadow; a second image appears slightly to the right—the "ghost" image.

The main image you see on your TV is the signal being transmitted directly from the TV station to your TV set. What causes this ghost is a second signal from the TV station that has been reflected, or bounced, off a tall building, or a mountain. Or, for short periods of time, even a passing airliner.

So this reflected signal reaches your TV a few thousandths of a second after the direct signal. That is why you see the ghost, or multipath image. The same thing is true of wireless networking. If you use a computer program called CommView, you will see, in search mode, that sometimes the same SSIDs appear on different channels. This may be multipath.

What can you do about this?

You can't change the way radio waves act, but you can use directional antennas and mount them on a tripod so that they can be aimed precisely where you want them. This may help improve packet capture on fairly weak signals. But then there is the example of the free Internet Café Soleil—it is three blocks away and in the opposite direction of where my grid antenna is pointing and yet I can detect the signal.

So, just accept that radio waves are somewhat unpredictable and wherever possible try to make line-of-sight connections.

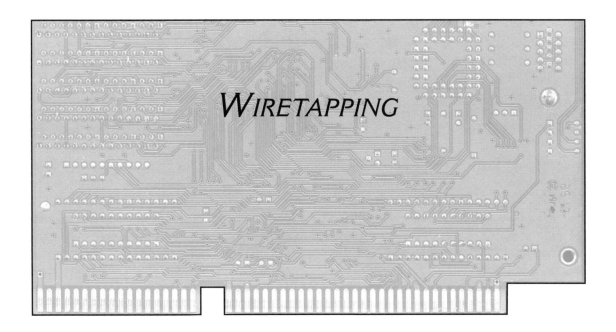

Wiretapping

Alexander Graham Bell is well known as the inventor of the telephone, but he was not alone in this project; a number of others were also working on the instrument. Elisha Gray, for one, had a working system, but as the story goes, Bell got to the patent office two hours before Mr. Gray.

The early phone was a wonderfully simple device compared to the systems of today. Wiretapping rules were simpler as well—there weren't any. It wasn't until the last decade of the 19th century, about 30 years after Dr. Bell is supposed to have spilled acid on his pants, that laws were passed making it illegal to listen in on people's phone line. But these were individual state laws; the federal government had not passed any such legislation.

In the 1920s Roy Olmstead, a rumrunner who imported alcohol from Canada during Prohibition, was arrested on the basis of information from a wiretap, and in 1928 the Supreme Court upheld his conviction. In his dissenting opinion, Justice Louis Brandeis wrote:

> The evil incident to invasion of the privacy of the telephone is far greater than that involved in tampering with the mails. Whenever a telephone line is tapped, the privacy of the persons at both ends of the line is invaded, and all conversations between them upon any subject, and although proper, confidential, and privileged, may be overheard. Moreover, the tapping of one man's telephone line involves the tapping of the telephone of every other person whom he may call, or who may call him. As a means of espionage, writs of assistance and general warrants are but puny instruments of tyranny and oppression when compared with wire tapping.

The Phone Book is about wiretapping as well as monitoring cell phones, with a little information on Echelon and other government surveillance systems thrown in; therefore, I will not attempt to duplicate the contents of that book here. I will make this a short chapter on the basics and what is new.

Now, then, it could have been on any of many places on the Internet where someone made a statement like, "Anytime you use your phone, assume that someone somewhere is listening," and "Even if you know *your* line is secure, what about the person on the other end?" Well, if you consider that hundreds of millions of phone calls are being made every day, you kind of wonder, how could the government or anyone listen in on all of them?

They can't. While the U.S. government would dearly love to be able to capture every word spoken by every person using a phone, this just isn't possible. They don't have the manpower and they don't have the money.

Do assume, however, that the feds are able to keep a record of every call you make, the date and time, the number you called or that called you, and how long the connection lasted. For eavesdropping on actual conversations, they have lists of the phone numbers of "persons of interest" that may be monitored in real time, but otherwise, I believe they will use various types of voice recognition systems that will scan wireless phone frequencies (with the technology to monitor spread spectrum) and microwave relay towers.

Now that there is VoIP (Voice over Internet Protocol), government surveillance such as Carnivore will be scanning the 'Net, searching for key words and phrases. Make a call and mention potassium nitrate and sulfur (two of the three ingredients of black gunpowder) and it may well be flagged and intercepted.

Thanks to CALEA (Communications Assistance for Law Enforcement Act), there are automated, hands-off intercepts without direct involvement by the telephone companies, and on a large scale. *All* forms of communication involving telephones in any way are subject to CALEA in the United States. Apparently, you cannot put up a new communications system with telephone connect unless you are CALEA compliant, at your own expense.

If the government taps my phone, is there any way I can find out about it?

Probably not; it depends upon how it is done. If it is a legal, court-ordered tap, it is very unlikely that you will ever know—at least not until after it is too late to worry. As far as I know, the Omnibus Act (Title 18 USC) still requires that anyone intercepted with legal surveillance equipment has to be notified by certified mail after the equipment has been removed—the infamous "wiretap letter." But the Patriot Act may have changed this.

The way they used do a legal tap, some years back, was with a little blue box in the frame room in the telephone company central office (CO). A phone call or fax from the feds to telephone company security was all that was needed for them to install it. Today, it is done electronically, and the phone company personnel don't necessarily even know the tap has been installed. And such a tap is impedance balanced, which means it does not affect the impedance of the phone line and doesn't make clicks or other noises; therefore, it is difficult to detect. No countermeasures equipment—not even the time domain reflectometer (TDR)—will detect it if it is in the CO.

There are places that advertise devices that they claim will detect such a tap. I have written to them asking for the loan of one of them to field test, but so far no one has responded. Matter of fact, very few suppliers of any kind of surveillance or countermeasures equipment have responded to my requests. A few actually refused to even sell the products to me when I advised them that I would publish the results of my thorough testing. That should tell you something. There are only a few companies that I know of (that deal with the general public) that I recommend, one of which is Shomer-Tec in Bellingham, Washington. Another deals mainly with law enforcement agencies, but there are a few exceptions. Contact me and I will maybe refer them to you.

As to "spy shops," be very careful dealing with them. First of all, none of them that I am aware of actually manufacture anything except hype. They sell equipment made by foreign factories, or REI, Shomer-Tec, Viking International, Capri Electronics, etc., and usually at a very high markup.

One remote possibility is if you are able to determine the phone company's information for your line—the cable and pair number—which is stored in a computer system that I believe is still called Cosmos, and then arrange for a tour of the CO. While inspecting the frame room with the thousands of pairs of wires, you might find your pair and see if something is there that is not on all the other lines.

If the tap is not a legal one, it isn't likely to be in the CO. So it will be in another place such as a bridging box (those big green cabinets you see on street corners and in alleys) or a junction point, or in your home or office. Should they place the tap inside your home or office, then the usual methods of searching, particularly the physical search, will find it. But this is rather unlikely. And it assumes, of course, that it is a physical, mechanical tap.

It is also possible to tap a phone by remote control by accessing the phone company's computer. Yes, that means exactly what you were afraid it might—that some people can dial up your line to intercept and record your conversations. More on this coming up.

As for wiretapping by non-law-enforcement agencies, private investigators, or individuals, this is problematic—they aren't likely to be able to place an impedance-balancing little blue box on the distribution frame at the local telephone company CO. Nor will it be easy for them to get into an underground junction point or find your pair of wires in a bridging box. But they may be able to access your phone line where it enters your home or office from outside, to the "demarc," or point of demarcation. Here, your pair of wires will connect to a junction box usually called a "66" from an old Bell system part number.

So, how does wiretapping work? There are, basically, three phases to tapping a phone.

- Finding and accessing the right line
- Making the actual connection
- Setting up the listening post (LP)

Making the connection is simple: Two transmitter wires are spliced to two phone line wires. Using alligator clips, this takes all of 10 seconds. However, getting access and then finding the right line may not be so easy; it may in fact be very difficult. Not many people can bug the White House, but Nixon did. Access …

Establishing the listening post can be done by stringing a pair of fine wires to a place where you can set up a tape recorder or by using a phone line transmitter that is capable of broadcasting its signal to a similar location. The range of such a transmitter depends on many factors, including power, type, and location of the antenna, and what is in between the transmitter and the receiver, such as walls and other buildings.

A few items you might consider before installing a tap are a good level-3 Kevlar vest and the funds for a good attorney in case the people shooting at you should miss. Seriously, this is a dangerous business, especially in the days of Homeland Security and all that.

HOW DO I SEARCH FOR A PHONE TAP?

First, picture in your mind what you are looking for: anything that is attached to the phone line—wires, splices wrapped in tape, any kind of little metal or plastic box, printed circuit board, anything that doesn't look like it belongs there. There are line-powered transmitters that are about the size of a pea. PK Elektronik makes one; another is the Russian Ruble. Also, look for an inductive tap. This will be a coil of wire of some kind, which may be disguised or placed in a plastic box or whatever. It will not be electrically connected to the line but will be very close to or touching it. It might be hidden inside something that the line is against or stapled to. A pair of wires will lead from the device to either a transmitter or tape recorder or to the listening post.

It is possible that you may not be able to access the entire line from the wall jack to the place where it leaves the building, demarc. And while it is generally true that if a wiretapper can get to the line, then so can you, there are exceptions, such as when a multiline cable is used and is tapped from a different apartment or office in the same building. In such a case, you can go to plan B.

The exact step-by-step process is in *The Bug Book*, but essentially, it works like this: Unplug anything that is connected to the line—phone, answering machine, modem, everything. Now go to the demarc and disconnect your line at that end. Measure the resistance with a good ohm meter. It should be infinite. Next, twist the wires together and go to the other end, your office or apartment, and measure it again. It should be very low, not more than about 10 ohms unless the distance is great—such as in a very tall building.

If the readings are not what they should be, then there is something attached to the line. Maybe it's a bug, maybe just an old AM radio interference filter (I found one of these some years ago), or whatever. At this point it is time for a more thorough investigation. Call in a pro that has a TDR or try to find a way to access parts of the line you couldn't before. Ever climb around inside an elevator shaft?

Now, back to remote-control wiretapping by accessing the telephone company computer.

Such technology does exist. Should you question this, please take a moment and think about it: The phone company runs on a computer using some version of the electronic switching system, and such a system absolutely has to be accessible from outside the CO for such things as "emergency" access by law enforcement. Suppose some important person is kidnapped, like a senator or a federal judge. Is the FBI gonna take the time to prepare the wiretap application, find a judge to rubber-stamp it, and then go to the CO and install the bridging tap?

Technicians in the field as well as within the CO have to be able to access lines for various purposes, such as testing and repair. With the right dial-up number and access codes, some people can dial in to the Electronic Switching System (or ESS, which maintains a record of every call made to and/or from a particular phone) and do all sorts of things. This was explained to me while taking a tour of a phone company switch (CO) in San Francisco. There is a photograph of such a phone in *The Phone Book*, which I keep shamelessly plugging.

This technique, sometimes called RemObs for remote observation, or SAS, exists. It is for real. I don't know that it works on all phone company switches in all areas, but it does work on the Northern Telcom DMS switches that much of the Bell System uses.

The way I understand it, SAS works like this: A computer with a modem dialup connection, a terminal emulator, connects to a secret number. Once contact is made, a series of login and pass code key words are entered, giving the caller access. They then enter the area code and number of the line to be monitored and call that line using a special telephone that listens only. The output feeds into one or more tape recorders as well as an amplified speaker. This may or may not make an audible click or a change in the line voltage.

Although they deny it, the FBI and other agencies will not rest until they have the capability of tapping any phone they want, from anywhere in the country, by entering the number into an automated computer system. Huge banks of digital recorders will make a permanent record of the intercepted calls, which will then be analyzed by special speech recognition systems such as the Cray processor in memory computers that are custom-built for the feds. And eventually, this *will* happen if it has not already.

I was once asked if it's true that the telephone company keeps a record of every call made on their electronic switching system—even local calls, calls that don't answer or were busy, etc. Yes, according to my guide at the CO I visited, this is true. The person asking me thought that the amount of information to be stored would be so great that this would not be possible. Let's look into this.

Consider the city of San Francisco. The population is a little under a million, but I'll round it up to that. Now, if every one of these million people places 10 calls a day (and that's probably a little high), that's 10 million calls.

Next, let's say there are 50,000 businesses, restaurants, retail shops, big corporations, and law firms in the city. And, averaged out, each makes 500 calls per day. That's 25 million. So, we have 35 million, including those cell phone calls that go through the CO.

Now, suppose the info stored for each call consists of:

- 50 bytes for the line originating the call including cable and pair number
- 50 bytes for each call made from or received on that line
- 25 bytes to indicate whether the call was answered, DA (no answer), BY (busy)
- 15 bytes for the time of the call

So, this comes to 140 bytes times 35 million calls, or about 4.9 gigabytes. Now suppose I am off by a factor of 10. So we have 49 gigabytes; that's not very much to store—my Western Digital 120 Gb drive would last two days, and racks of dozens of similar drives would last for months before being stored and replaced. And, according to someone at Western Digital, info on an unused (stored) drive may last five years or more.

WHAT IS A PEN REGISTER?

Although they are not all exactly the same thing, a pen register is also called a "trap" or "trap and trace" or DNR (dialed number recorder). It is a function of the phone company's ESS. Since it does not intercept a wire or oral conversation, it is not illegal for law enforcement to use one without a warrant. Apparently, they just fill out a subpoena form and take it to phone company security and it is done.

Can it be defeated?

AT&T and perhaps others have a service where you call information to get a number and then are told that for a small fee, like 50 cents, you can be connected. Because the phone company is dialing the number and you are not, all the pen register will record is that you called directory assistance. In some areas, you may not hear this message unless you stay on the line after directory assistance disconnects. Try it and see.

Also, there are call forwarding services where, for a small fee, you can dial them, get a second dial tone, and then make the call from their location.

The easiest way to avoid this intrusion is to use a pay phone.

ANSWERING MACHINES

Most of the machines on today's market have a remote control feature—call it from anywhere and retrieve your messages. Convenient, but dangerous. All of the machines that I know of use a simple three-digit access code or password. There are inexpensive devices available with which a person can access your machine, listen to and delete your messages, and even change your greeting—perhaps to something that your callers would find offensive. There is, however, an inexpensive product that counters this and makes your machine secure against these invading devices. Details, sources are in *The Phone Book*.

THE DROP-OUT RELAY

This is a simple device that, when connected to a phone line, will automatically turn on a tape recorder, capture both sides of the conversation, and turn off when the phone is hung up. They are available from a number of sources, and Radio Shack may still have them.

There was a time when, if someone hinted that the government was tapping everyone's telephones, that person was dismissed as a wacko, paranoid, in need of a tin-foil helmet and some Thorazine. But not any more. Next, a story based on my experiences in the early days of telephones, which I

> WHETHER OR NOT THIS INCLUDED THE PARTY LINES that in some areas still existed into the 1960s, I don't know, but it didn't keep some people from listening to neighborhood gossip.
>
> Back then, I lived in an apartment in the Midwest where there were party lines. Two of the people on our line were an old biddy in the house next door who loved to gossip and spread rumors about anything she could think of, and a girl in the upstairs apartment who loved to listen and interject a few thoughts of her own. But when you picked up the receiver, it made, usually, an audible click so whoever was on the line knew that someone was listening in.
>
> So I somehow came up with an idea. There were little two-transistor battery-powered audio amplifiers available at the local radio parts store, called Barr Electronics. I connected one of them to the phone line through two capacitors, and whenever anyone used the party line I could hear them without them knowing.
>
> Ah, the juicy and sometimes bizarre things I heard.
>
> This led to the Listen Down amplifier you can read about in *The Phone Book* and easily build yourself!

hope will elicit a few chuckles, and then on to some other surveillance methods.

THE "TELE-PHONE"

Once upon a time, when I was a kid, not everyone had a tele-phone.

Matter of fact, a lot of people didn't. Our family didn't, and neither did most of our neighbors, so it wasn't a big deal since we didn't have anyone to call anyway. But we had our networks before anyone even thought of the term. There was the Grange Hall, the cracker barrel down to Fennerman's General Store and Mercantile. For the grown-ups there was the First Chance/Last Chance Tavern where I used to sell the empty pop bottles I collected in my Radio Flyer wagon, the ladies caught up on (or more likely started) the latest gossip at Estella's Tonsorial Parlor and Beauty Emporium, and the old folks had the IOOF, which wasn't exactly the same as the Odd Fellows lodge.

Oh, Fennerman had a "business" tele-phone, which he always explained it as being, whenever someone wanted to use it. He took one of the spittoons and put it on the counter so that everyone who wanted to try it had to drop in a nickel. Funny, how folks quit askin'.

But well, most of us got along quite well without one a them newfangled gizmos.

It wasn't long before these here fellers came along all the streets a-stringin' up these wires and then hookin' 'em up to these little metal boxes on folks' houses. And then this city slicker in a Sears-Roebuck catalog suit come round, a-tellin' everyone how important it was to comm-yun-e-cate. Shucks, he weren't tellin' us anything we didn't already know, 'taint like there was any secrets in a little town in southern Ohio in the 1940s. 'Sides, if it were something of import, Walter Wolf, the police chief, had a tele-phone and it didn't take long to run on down there to the jail and wake him up.

Now, my daddy, bein' a man of the cloth, decided that he should have one a them tele-phones put in the parlor since he had to minister to the needs of the infirm and give them Communion in their homes, 'cause they couldn't make it to the church. Now, my daddy never did figure out why some of his parishioners wanted communion so often. Me and my big sister thought it was funny—ya see, my daddy didn't believe in spirits ('cept the Holy ones he was always preachin' about) so he used grape juice for Communion (till we started fillin' his silver flask with some of the stuff we found in bottles behind the First Chance/Last Chance. But that's another story).

And so, we got us a tele-phone. Made out of oak and was nailed to the parlor wall. Had this nifty crank that you turned to get the operator. Our ring was two longs and one short. But me and my sister always picked up the ear-thing just to see who was talking to who. Which, amidst my reminiscing, brings up a point. Back then, folks answered their tele-phones, even if it was just to eavesdrop. "Hello" was soon to become a household word.

Well, one day, in the mid-1950s, my daddy moved us to a very big city where we lived in a very small house. And, not only did it have a tele-phone, we even had our own number: Eastbrook 657. The phone was one of those modern kind where you didn't need to crank up a cranky (the stories I could tell you) operator. It has this nifty round thing with letters and numbers on it. Turn it just the right number of times and you could comm-yun-e-cate with folks all over. Soon as it started making these buzzing sounds, sure enough, someone would say, "Hello?"

Yup. Folks actually answered their tele-phones.

I was fascinated with this, the latest in communications technology, and even though it was absolutely forbidden (Ma Bell was more to be feared than even Senator McCarthy), it wasn't long before I took it apart. I just wanted to see what was inside it. How it worked. Well, the dreaded Phone Company didn't understand my innocent curiosity and I was sent to the Sol A. Wood Juvenile Detention Center. There, I was told that I should resist my evil ways, that communicating with The Lord should be the most important thing in my life.

Funny, though, as they never did give me His number.

I just couldn't contain my curiosity about tele-phones and so I learned everything I could about that secretive brick building where all the lines were connected together with Strowger switches. And that the back door wasn't always locked. And oh! The things I found in the trash cans! Yes, I was one of the original phone phreaks and a regular at Sol's Place.

Ah, well, time marches on and progress (?) follows. And in the late 1950s, something happened that would forever change what Ma Bell had in mind—to use the tele-phone to comm-yun-e-cate. I had just set up one of the first speaker (hands-free) tele-phones using parts stolen from Superior Scrap Iron and the Western Electric junk yard in the dead of night. Some of the ham operators heard about it and wanted to come over and see it work. (So they said. Actually it was to mooch my stock of Carling's Black Label and Cheez-Its.)

So, I spun the black dial, and this is what I got:

"This is an electronic secretary, automatically answering the tele-phone for Acme Window Cleaning." (Please leave the proverbial message after the infernal beep). Aaarrgghhh.

And so began an era where people do not answer their phones. After all, in this hectic society, folks cannot be expected to sit by their desks waiting for someone to call. America was becoming mobile.

In the mid-1980s, the cellular tele-phone system became operational. People could carry their tele-phones with them, so that they could answer—say "Hello"—no matter where they were. But, for some reason, they still do not. You want to call your Granny to see if she got the birthday card you sent.

Click, click … ring ring … and what do you get?

"Hello"?

Nope.

"Welcome to the AT&T wireless voice messaging system … If you know your party's extension … "

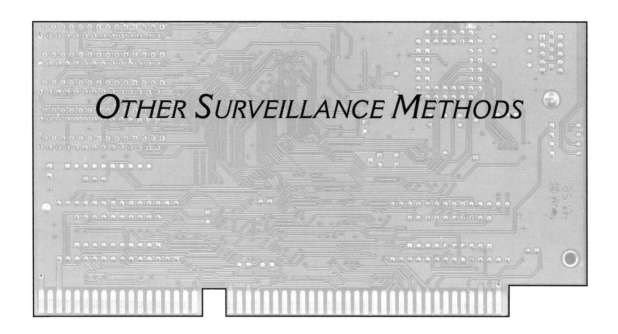

OTHER SURVEILLANCE METHODS

VEHICLE TRACKING

Remember the scene in *Goldfinger* where 007 has the magnetic transmitting device on Goldfinger's car, and the agents plot the location on the screen of a tracking device in their '62 T-Bird? When that movie was produced, such technology did not exist. Today, it does. There are many systems that use Global Positioning System (GPS) technology and, on a local level, LoJack to determine the location of a tracked vehicle.

A few years ago, some people from back East where I used to live came to visit. I met them at the San Francisco airport where they rented a Cadillac SUV. It was equipped with some kind of system, probably General Motors OnStar, where the driver could call up a dispatcher in case he needed directions. But what is important here is that the Caddy was being tracked. Wherever it went, this system knew its location.

The system worked well (which, incidentally, is more than can be said for the people on the other end providing directions. The human that answered took them far out of their way).

If you have a late-model vehicle and you have this system installed, then it is possible for the feds to know precisely where you are at any given time.

Some vehicles have installed an anti-theft system called LoJack. If your car is stolen, you can report it and the LoJack system activates the homing transmitter so that law enforcement agencies can track it and recover it. Well, that's all well and good, but since the transmitter is activated by remote control, you have no way of turning it off. Any police agency—in particular the feds—can activate the transmitter whenever they want. Wherever you go, you are being tracked.

These tracking devices are available to the public at many shops listed in the Yellow Pages under "vehicle alarm systems." They are installed while you are waiting outside, looking at ragged copies of *Mechanics Illustrated* and *Popular Pistons* and perhaps not even realizing that you don't know specifically where in the vehicle the equipment is being installed. You aren't allowed to know, but one person at such a shop said that there are up to 90 different places where they might be hidden inside a typical vehicle. They may also be installed by federal agents or Industrial Strength Spies, but they would have to get uninterrupted access to the target car.

Another type of vehicle tracking device, often called a "bumper beeper," has a fairly small range—perhaps a block or two, maybe miles—depending on many particulars. Usually battery operated, it may be placed under the vehicle, which is a science in itself. See coauthor Steve Uhrig's Web site at http://www.swssec.com and read "A Day in the Life" and some interesting articles on vehicle tracking.

Yet another type of tracking that probably isn't often thought of as such is the increasing use of devices that are installed in vehicles to automatically record their entering a toll way or bridge. Convenient, yes, because the driver does not have to stop to toss a coin in the hopper, but remember—this is also just one more way that the government can keep track of you, and where you go, and when you went …

What are some other surveillance methods?

NATURAL AUDIO PATHS

Surveillance can happen virtually anywhere and it doesn't have to be high-tech. It can be low-tech or even no-tech. A water glass with the bottom pressed against a wall might make it possible to hear what is happening on the other side. A physician's stethoscope works even better.

In homes that have heating ducts, if you open the registers in a room you want to monitor, you can often hear people talking. And using the stethoscope again makes the sounds even clearer.

The electrical conduit that carries wires to wall plugs might reveal people's conversations if a small microphone is placed inside. But remember that there is 117 volts there and that is enough to kill a person who isn't careful.

FLICKERING LAMPS

There are devices that can be built into a table lamp that use room audio to modulate or vary the brightness of the bulb, but not to the extent that anyone would notice. Meanwhile, someone in a nearby vehicle with equipment similar to that used in laser systems can capture this audio.

An anonymous person sends you a nifty lamp? Maybe time to be a little suspicious.

RF FLOODING

This is a method of beaming a strong RF radio signal at a multiline telephone. Sometimes this will cause conversations on the same system to emanate from the speakerphone in other offices.

Installing a vehicle tracking system

The Tracker

SUBCARRIER TRANSMITTERS

Wireless intercoms that send their signal through the power lines can be intercepted using the same type of intercom or special types that hear all subcarrier frequencies.

Microwave systems (both outside and inside) were used against the U.S. embassy in Moscow some years ago. Now, this is fascinating: One such system was the "Great Seal," which made headlines some years back and is described by Glenn Whidden in his book *The Russian Eavesdropping Threat – Late 1993*. "The device was a passive cavity resonator that was concealed in a wooden plaque that has the seal of the United States carved on its front surface. It was presented as a gift from the Soviets to the American Ambassador at Moscow. It was powered by a radio signal that was transmitted to it from a point outside the embassy and radiated a radio signal that carried room sounds from the Ambassador's office."

And of course when the building that housed the U.S. embassy in Moscow was built, the steel reinforcing bars were made into the shape of a parabolic reflector similar to those used with directional microphones. Audio from inside the building caused tiny vibrations in the bars, which caused variations in the reflection of a microwave beam. This was converted into audio, similar to the laser devices described above. I don't know how well it worked, but using microwaves would have at least eliminated some of the problems with visible laser light, such as rain and fog.

Spying in Real Life:
Surveillance Installations

by Steve Uhrig

Note: The pictures in this section were taken during actual law enforcement surveillance operations using equipment built and/or installed by coauthor Steve Uhrig.

Although my company, SWS Security, works quite a bit with private detectives, governments around the world are our primary focus. Recently we manufactured and installed the nationwide pager monitoring network (our Beeper Buster system) to antinarcotics forces in Bogota, Colombia. This system allows the government to monitor any message sent to any pager anywhere in the country. Several years ago, the Colombian military used our pager intercept system as a primary tool to dismantle the second biggest narco-trafficking ring in the history of the country.

We also manufactured and installed a multimillion-dollar communications intercept platform to the U.S. Army on a contract to Bogota, used to intercept, unscramble, record, and analyze radio transmissions by the drug cartels in the jungles. This network consists of 44 systems installed in locations around the country.

Also in a recent case, SWS responded to an emergency request for assistance from a Central American government on a kidnapping of a high government official. In full panic mode with only a few hours to catch a plane, we threw together radio tracking beacons (transmitters) and a direction-finding receiver.

The government tasked us with bugging the ransom money so we could trace it to its destination and hopefully recover the victim. It was rather interesting hiding the transmitter in a stack of bills. We installed direction finders and allied electronics in several ground vehicles and airplanes and were on standby. Unfortunately, after several days of standby, we learned that the victim had been killed in an exchange of gunfire between the bad guys and the victim's security team when he was taken. So we never deployed, but it was valuable experience for the next time.

And there will be a next time, someday, somewhere.

Even though the drug dealers were arrested, the technical details of how this surveillance was established are still classified because the case is still working its way through the court system.

These photos were taken in a hotel room next to the one where some big-time drug dealers had planned a meeting. It had been set up some time in advance, so the law enforcement agencies had time to install their surveillance equipment here.

Ozzie Eans of CSE Associates working a surveillance operation in South America.

Training antinarcotics agents in SWS- and CSE-manufactured communications intercept equipment at a secret location in Bogota.

Installing video and other monitoring equipment in Embassy Row in Washington, D.C.

Part Three

ELECTRONIC COUNTERMEASURES

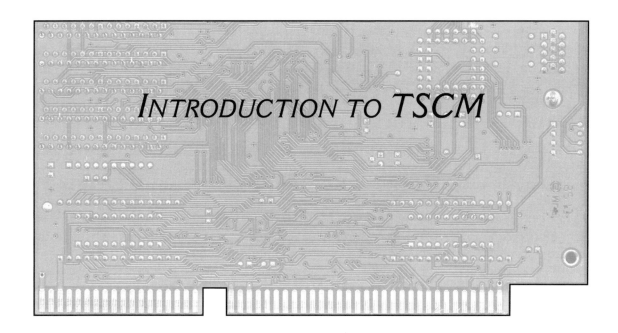

Introduction to TSCM

IMAGINE YOU ARE CHIEF SECURITY OFFICER FOR A BIG CORPORATION AND YOUR JOB is to make the premises secure from any form of surveillance—to ensure that no one has placed any kind of listening device in any of the offices, labs, or anywhere on company property. You use your connections and hire a team of countermeasures experts who sweep the entire area using the latest technology, including thermal imaging, but are unable to find any spy devices.

Does having a search done by experts absolutely guarantee that the area is 100 percent clean—that there is no way anyone could be eavesdropping on you?

Not exactly. Not necessarily. All it means is that the technical surveillance countermeasures (TSCM) experts have made a diligent search and based upon that, their experience, and their equipment, they were not able to find any listening devices. Again, remember that this has to be done within reason. Leveling the building would provide this absolute guarantee, anything less than that might not. In between the extremes are some things to be considered, to maybe be done.

Consider a very old building that has been vacant for years and has recently been rented by a business. It is reasonable to assume that the batteries for a surveillance transmitter would have long ago gone dead; however, a device connected to the lighting circuit might well still be active. What's important here is that while the battery-powered device might be, probably would be, much more difficult to locate physically, anything connected to the power lines would not be so difficult to find. Just follow the wires.

The same is true of telephone transmitters. Trace the wires. These are the things an expert countermeasures technician has to know and understand and apply to the job.

OK, so who are these "experts"? Where can I find them? How will I know if they really know the TSCM business?

If you really want to know the answers to these questions, then consider getting a copy of *The Phone Book* before you spend several thousand dollars for a sweep. It has a detailed chapter on countermeasures and a series of photographs of equipment as it is actually set up on a sweep. These are not stock pictures from the manufacturers or taken from a catalogue; I took them myself with my trusty Nikon while I was working a sweep. Included are the Eagle Plus, Scanlock, a spectrum analyzer, Tektronix TDR, and a spectrum monitor, among others. The book also includes an equipment list, a description of an actual sweep, and more. You will learn about the gear, terminology, procedure, methodology, reports, etc., and with that information you will absolutely be able to tell the experts from the amateurs, phonies, charlatans, and wannabes who vastly outnumber them.

Am I saying that more than half, 50 percent, of those who claim to be experts in the countermeasures business actually are not?

Yes. That is exactly what I am saying.

There are maybe a couple dozen real experts in the entire country. So, if you are seriously considering a sweep and don't know who to hire, e-mail me and I might arrange to have you contacted. Might. Most professional TSCM teams will not do sweeps for individuals in private homes, but there are some exceptions. When I get letters asking for a referral, I consider the content and then may pass it along to one of the team leaders I know. It is up to them if they contact you.

So, back to what you can do to protect yourself and your family against eavesdroppers.

Get *The Bug Book* available at Paladin and some local stores such as Barnes and Noble. Read it carefully and use the sample checklist in the back as a guide to making your own physical search.

Then, once you are convinced the area is secure, use some of the other techniques.

A few examples:

- Seal the plates on switches and plugs with two colors of nail polish, or get some transparent UV sealer from Shomer-Tech.
- There are tamper-proof adhesive labels that can be applied to virtually anything that opens or can be opened, such as electronic equipment and appliances (one bug was found in a coffee maker as you can read about in a coming chapter).
- Get ID from anyone who claims to be a maintenance person from the power or phone company or whatever. Write down his name and advise that he will be videotaped as long as he is on your property.

As to a do-it-yourself sweep, you might want to skip ahead to the chapter on becoming a counterspy and then come back to read the rest of this chapter

It is not intended to dissuade you, only to point out that to do the job right requires a great deal of experience and, of course, a lot of expensive equipment.

Again, I suggest you read *The Bug Book*. But when you get started finding RF transmitters, avoid cheap devices that claim to find "any hidden transmitters on the premises." Don't waste your money on false hopes as there is a great deal of overpriced junk out there, the sellers of which will be happy to take your money.

Some medium-priced equipment, all of which is of very high quality, is made by Marty Kaiser. And though he is retired, some of it is still available if you search for it. Many pictures, descriptions, and specifications are in *The Phone Book*.

Now, as to bugs, the vast majority of homemade types, commercial wireless microphones, and those sold by "spy" shops will operate/transmit on frequencies that are covered by many of the police scanners on the market, as well as communications receivers. So if you have such a radio, or can get one, use it and tune through its entire coverage with the volume turned up. Way up. Move the radio around, set it up in all of the rooms in your home or office and don't overlook the kitchen and bathroom. (Ah, the stories from my experiences that I cannot tell you.)

Time consuming, isn't it? So what if there were a way to get the radio to do this automatically? Many radios can be computer controlled—programmed to sweep through whatever frequency range they cover.

If you are lucky, you might find a Realistic PRO-2006 with the Optoelectronics computer control printed circuit board installed. It covers the spectrum from 25 to about 1200 MHz (including cellular telephone channels), where the vast majority of bugs and wireless microphones transmit.

Many other radios are available that can be computer controlled. Do a little research and you will see many listings, but before you buy, you might read the specifications. Some scanners skip important bands that may be used for bugs. The Realistic trunking and conventional scanner PRO-95, for example, does not cover 54–108 MHz, where many wireless transmitters operate.

Now as to computer control, or CAS (computer-aided scanning), there are several programs that work with various radios. I have tested some of them, and for countersurveillance as well as just general listening, the one I use (on an ICOM R-8500 communications receiver) and highly recommend is Radio Max, available from http://www.datadeliverydevices.com.

It's an excellent program—versatile, powerful, and inexpensive.

Next, a look at thermal imaging, something that only a few countermeasures teams have.

Seeing Red with Kevin Murray: Infrared Thermal Imaging to Locate Surveillance Devices

Copyright © 2002 by Kevin D. Murray. Reprinted with permission.

THERMAL EMISSIONS SPECTRUM ANALYSIS® (TESA) IS JUST ONE THE MANY techniques developed by Murray Associates to enhance the speed and effectiveness of surveillance detection. Murray Associates is recognized as the first firm—and still one of the only firms—to make this advanced detection technology generally available to private sector clients. The following is from their white paper on the subject, *Thermal Emissions Spectrum Analysis (TESA) as a Detection Technique for Finding Covert Electronic Surveillance Devices* by Kevin D. Murray, CPP, CFE, BCFE, and board member of the International Association of Professional Security Consultants.

Heat is the graveyard of electricity; it is where expended electrons go to die. Look in the right places for these graveyards and start digging—you just might find buried electronic surveillance devices: audio bugs, microsize video cameras, recorders, wiretaps, and the transmitters that move private sights and sounds to illicit eyes and ears.

The premise is simple: When electricity moves through any electronic circuit, some of the energy converts to heat. This is caused by resistance, which is inherent in all circuits. Cooling a circuit to a temperature of absolute zero (0 degrees Kelvin / -45 degrees Fahrenheit) is the only way to eliminate resistance. Fortunately, refrigeration of electronic circuits is not practical in the real world or in the shadowy world of espionage. Electrons will meet with resistance: Heat will be generated, heat will migrate, and heat can be detected.

Heat may be generally thought of as light waves that are too low in frequency for our eyes to see; thus the term used to describe this is "infrared" (below red). Neither can we hear radio waves, dog whistles, and bats echo-locating sounds because the frequencies are either too high or too low for our ears to hear. To perceive these out-of-our-perception frequencies we need some type of instrument that will detect and then convert them into something we can perceive. A thermal imager is basically a converter—taking low-frequency light waves and converting them to the light frequencies we can see. Loosely speaking, radios perform the same function for our ears.

The amount of heat radiated by a bugging device is quite small. The sensitivity measurement required to see this is in the order of eighteen-thousandths of a degree Celsius or less! By comparison, human touch can distinguish a temperature difference of only about two degrees Celsius by most accounts.

TSCM practitioners have suspected for a long time that heat sensing would be an aid in surveillance device detection. Unfortunately, portable instrumentation of sufficient sensitivity and resolution was not available until now. The first and most common imagers—Level I—were not effective enough to be useful for bug detection work.

SCIENCE CATCHES UP TO OUR DREAMS

Level II imagers are now available.

These instruments are truly useful. Currently, price is the only thing standing in the way of complete acceptance and adoption to TSCM toolkits. The good news is that rapid advancements in the field of thermal imaging sensors, combined with increased demand, should cause prices to fall into acceptable ranges.

ADVANTAGES OF THERMAL EMISSIONS SPECTRUM ANALYSIS

There are three stunning advantages that a Level II thermal imager brings to a TSCM inspection:

The ability to see slightly elevated temperatures. Example: It can inspect 40,000 square feet of ceiling tiles in less than 5 minutes to find a 1-inch square video camera embedded in one tile.

The ability to see density differences in materials. Example: It can inspect 40,000 square feet of ceiling tiles in less than 5 minutes and find where a video camera was—at one time—embedded in a ceiling tile.

The ability to see through some materials. Example: Certain materials that appear opaque in normal light become quite transparent when viewed at infrared wavelengths.

TESA is not a replacement for any other TSCM inspection procedure. It is an additional technique that greatly enhances effectiveness and will allow for the most thorough inspection possible when the time allocated for the inspection is limited.

ABOUT THE AUTHOR

Kevin D. Murray has been solving eavesdropping and espionage concerns for business and government since 1973. He is also credited with several important technical advances in the field of eavesdropping detection, which can be seen at http://spybusters.com/.

Murray is frequently quoted by well-known media sources such as *Fortune* magazine, the *New York Times*, *USA Today*, *National Public Radio*, *International Security News*, *Corporate Security*, *Security Management*, *Congressional Quarterly*, *Security Letter*, *Time*, *Boardroom Reports*, the *Washington Post*, *Business Finance*, and several Internet security sites.

His other published works include *Kevin's Security Scrapbook*, which is his weekly spy news e-letter (available at his Web site).

Murray's electronic eavesdropping detection and counterespionage consulting services are available throughout North America and selected foreign countries. He considers writing and teaching a necessary evil. His true love is developing new detection technologies and working for his client family to keep them safe from espionage attacks.

Copyright © 2002 by Kevin D. Murray/Murray Associates. All rights reserved. Thermal Emissions Spectrum Analysis and TESA are registered trademarks of Murray Associates, http://spybusters.com/.

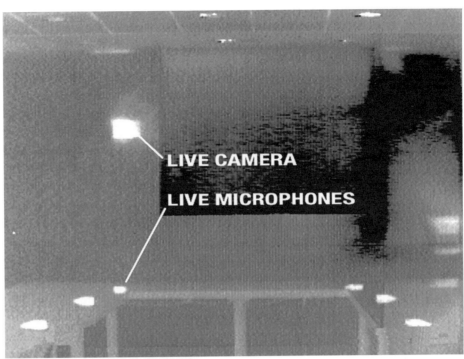

Far left: The thermal imaging system in action. It's not much larger than a camcorder.

Above: Hotspots show the location of surveillance devices—camera and microphones.

Left: Thermal imaging of a ceiling tile.

Lower left: The devices uncovered.

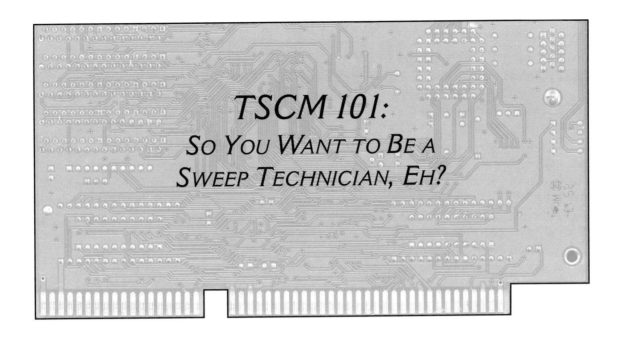

TSCM 101:
So You Want to Be a Sweep Technician, Eh?

SINCE YOU ARE READING THIS BOOK, YOU MUST HAVE AN INTEREST IN surveillance and countermeasures. So what if you wanted to make a career of it?

Getting a job as a countermeasures operative isn't like most other professions. You could search the want ads in a hundred newspapers, or the online listings at employment agencies, but it is unlikely that you will ever see an opening for a tech. It is a kind of closed circle where most of the people in the business know each other and any openings are likely to be word of mouth. So, I will not offer you false encouragement. I will not imply that this is an easy profession to get into. It is not.

It's true that since the terrorist attacks of September 11, 2001, this is changing. There is a growing need for people, capable, knowledgeable countermeasures technicians, and I believe this will continue to increase. But still, how can you break into this fascinating and important profession if you have no experience?

Learning to be a good countermeasures technician starts with the fundamentals, with reading what some of the experts—the legends—have written. Learning means trying to get inside their minds, understand how they think, and know what they are trying to say in what they have written. In Appendix B I have listed some of the classic books on surveillance and countermeasures. Many of them are out of print, but if you are determined, you will be able to find them.

You will learn from reading *The Bug Book* what a surveillance transmitter looks like. (It can look like virtually anything.) You will learn about wiretapping from *The Phone Book*. Do a Google search for Glenn H. Whidden and study—don't just read—his books. Whidden is an expert in the field of surveillance and inventor of some excellent countermeasures equipment.

Also, log on to SWS Security's Web site at http://www.swssec.com and read *everything*, especially the White Papers section, keeping in mind that SWS does not normally make referrals for private residences.

Now, let's look at aptitude and at developing the attitude—the mind-set, the way of thinking. You need to learn to know what to look for, what doesn't belong. You must automatically think when you walk into a room, "How would I bug this place?

How would someone else?" It is a kind of second nature feeling, an ability to sense things—"This doesn't look right" or "That doesn't belong there."

A good sweep technician will be a do-it-yourselfer—someone who is mechanically inclined; who doesn't need to call in a plumber to fix a leaky faucet; who works with his or her hands. Some knowledge of electronics is a big plus in getting started. If you have built things or worked on circuit boards with a soldering iron, you are familiar with the flux residue that results from a hot iron connecting things. This yellow-brown crystalline substance drips off the wires being soldered; if you are examining a telephone and see signs of this residue, you are instantly alerted since telephones do not normally have soldered connections that leave this flux.

When you pull the cover from a power outlet, you already know that there should be a white wire, a black wire, and a bare copper ground wire, and nothing more. If you see any other wires, or a tiny little piece of fiberglass with small wires on it, you instantly know this is a possible listening device.

These are the some of the skills and observational abilities that are the makings of a good countermeasures technician.

Obtain and read the manuals for all of the countermeasures equipment you can get your hands on. Learn what this equipment does. What is a nonlinear junction detector? What does it look like? How does it work? What, exactly, does it do? Learn all you can about TDRs and spectrum analyzers. Even if you have never seen any of these instruments, just knowing what they are and how they work will make a better impression than knowing nothing about them.

Having a ham license is a plus because it shows you are familiar with radio equipment. The majority of countermeasures techs are licensed amateur radio operators

If you have, and know how to use, a good quality communications receiver, this is another plus. Sweep teams use these radios, such as the ICOM R-8500, to follow up on signals intercepted by the Eagle and Oscor countermeasures receivers.

Finally, a good understanding of wireless computer networks is one of the most valuable skills you can have, since this is becoming a wireless world. If you have the aptitude and the determination and can network with the right people, you might break into an occupation that is exciting, but damn hard work. And after you have several years of experience and can raise a hundred thousand dollars or so for equipment, you might even start your own company.

THE BUSHWHACKER™
FROM CSE ASSOCIATES

THE BUSHWHACKER IS A VERY SOPHISTICATED (AND EXPENSIVE) RADIO direction finding and mapping system with which it is possible to tune in on a radio transmission, such as someone jamming a law enforcement channel, and pinpoint his location on the built-in maps. And it does this from a single physical location.

Traditionally, triangulation required two base locations using direction-finding gear. Each determined the direction the hunted signal was coming from and then by converging both on a map, the approximate location could be determined.

Bushwhacker uses something called an "Adcock" array—which consists of five small vertical antennas. These antennas are electronically rotated, turned on and off in sequence so that the signal can be seen as if from different places.

Then, the location of the hunted transmitter can be plotted on the map as seen in the photo below.

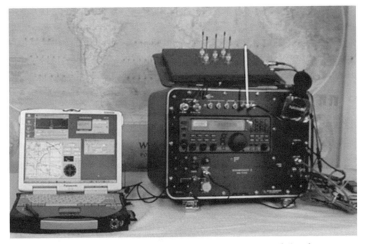

The "Adcock" array on the Bushwhacker consists of the five small, vertical antennas seen at the top of this photo

The map shows the location of the hunted transmitter.

A very sophisticated system indeed, but since the Icom R-8500 radio it is built around is unblocked and can tune cellular radio frequencies, Bushwhacker is for sale only to law enforcement agencies.

So, between that and the cost, I am unlikely to own one (which is too bad, as I'd love to find the jerk that interferes with transmissions on our local Amateur Radio Emergency Service net on 443.100).

Next, let's go over what might or might not be a bug.

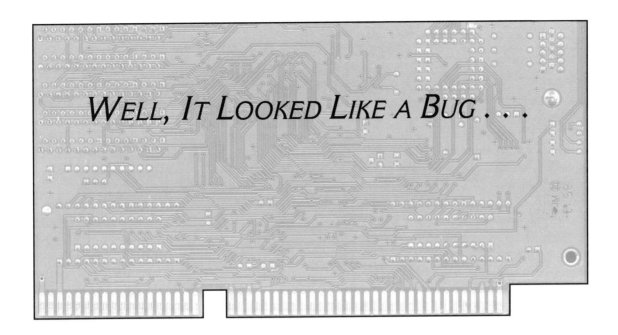

Well, It Looked Like a Bug....

In *The Bug Book*, I wrote about two incidents in which someone found what he believed was a hidden listening device. Neither of the gizmos turned out to be bugs, but I used these examples as part of the learning process. Here is an example—a real-life true story—from coauthor Steve Uhrig of how a professional technical analysis of a suspected listening device is conducted, as well as the report format.

> 3 November, 2003
> (Name and address of sender sanitized)
> Via courier
> Sensitive
> Re: Unknown/suspected eavesdropping device
> Dear (Sanitized),
> I have examined the suspect device you shipped to me.

Background—The device reportedly was discovered during cleanup of a conference room in a repossessed facility owned by a major international corporation. This corporation presumably would have held and processed information of interest to competitors, and also is involved in union dealings. There is a suspicion organized crime was involved in the operation of the parent company's distributorship, which managed the facility where the instant device was discovered.

The device was discovered magnetically attached to a metal rectangular-shaped serving tray. The metal tray is manufactured of stamped, light-gauge ferrous sheet metal painted with a multicolored advertisement for the company and one of its products. The tray measures approximately 7.0 inches by 4.5 inches. An inscription on the bottom of the tray indicates it is an official licensed product of the parent corporation and bears a copyright logo and date of 1992. This tray was located on top of a cabinet in a conference room. Reportedly, the custodial crew that discovered the device turned it and the tray over to the parent company's corporate security director, who turned it over to you in your official capacity with the government. You in turn contracted me and conveyed the device to me for analysis. This is information as given to me; I cannot vouch for precise details or accuracy, only a general background.

General description—The device presents as a small, metallic, flat, round object of some complexity and quality. At first glance, it does appear to be an electronic component or device.

The device measures 0.486 inches (12.34 mm) in diameter, 0.065 inches (1.65 mm) thick, and has two raised, spring-loaded terminals for electrical connection on one large flat side. The thickness of the device including these terminals is approximately 0.106 inches (2.7 mm). I say "approximately" because the terminals are spring-loaded and compress when measured with a micrometer. Metric measurements are in parentheses.

The device weighs 1.134 grams.

There are a series of characters printed in ink around the perimeter of the device. This string of characters is: 12345ABCDE 24680 FGHJK

These characters are printed on what I will refer to as the bottom of the device, which is the side with the electrical terminals.

I have not included photographs of the device in this report as the ones you provided likely are as good as anything I could do here without an elaborate macro photography setup. The device incorporates a fairly strong magnet. The magnet is a cylinder or pellet recessed within a machined well in the device. The well actually is machined all the way through the device at a constant diameter. Picture a slice of metallic tubing in the given diameters and that is the raw configuration of the slug. The metallic slug is cast or machined though, manufactured with some quality. It is not a simple stamping.

The diameter of the magnet is 0.065 inches (1.65 mm). I cannot measure the thickness of the magnet absent a destructive disassembly of the device, and for our purposes it does not matter. The magnet appears from physical inspection and its properties to be a rare earth magnet, probably samarium-cobalt or neodymium. The color is a very dark gray, essentially black.

The "top" of the device, i.e., the large flat side opposite the terminals, is covered with a transparent flexible plastic membrane.

This membrane is glued around the inside perimeter of the metallic housing, with a small amount of overlap, meaning the diameter of the membrane is slightly larger than the diameter of the device itself.

This membrane is slightly peeled off, probably from handling prior to my taking possession. I took great care to prevent it from peeling any farther.

Analysis—Upon inspection under fairly potent optics, I notice a coil of very fine enamel-insulated wire, perhaps AWG 40 or finer—in other words, extremely fine, or probably finer than a human hair. There may be 50 or more turns; I am unable to determine absent disassembly of the device.

The coil is wound around the magnet and the edge of the coil is glued to the membrane.

The two wire ends of the coil plainly are soldered to the two spring-loaded contacts mentioned earlier. The bottom of these contacts and solder joints are visible through the transparent membrane.

Conclusion—The device is a dynamic microphone element. A dynamic microphone element consists of a diaphragm coupled to a coil, with the coil surrounding the magnet. Sound pressure acoustically vibrates the diaphragm, thus moving the coil in and out in correspondence with the sound pressure.

The movement of the coil generates a current due to the mechanism being an electromagnet.

The current is a reasonably accurate representation of the audio intelligence in the vicinity.

The device converts sound to electricity.

Dynamic microphones can have various patterns such as omnidirectional, cardioid, etc. The pattern is determined by the physical packaging of the element as the primary variable. Since we have the element unpackaged, we cannot determine the pattern for which the microphone was intended.

To prove the theory of the device being a microphone element, I clamped the device in a rubber-jawed electronics vice. By touching the tiny spring-loaded contacts with needle probes connected to an oscilloscope, I was able to see an audio AC output on an oscilloscope while whistling at the device.

The DC resistance of the element's coil measured at 121.55 ohms. Normally, dynamic microphones are low impedance; however, this impedance can be transformed via external transformers.

I did not take photographs of the test setup nor record any parameters or measurements, as I considered them immaterial to this report. Nor did I make any attempt to determine the frequency response of the microphone element. However, the characteristics of a dynamic mic element are well known. In general, they have a frequency response rather similar to the human ear, have relatively low outputs generally in the single-digit millivolts, and are moderately but not extremely sensitive. Note: The sensitivity of a microphone element depends on, amongst other parameters, the mass of the diaphragm needing to be modulated with the available sound pressure.

A representative output spec for a similar mic element is: 2mV/Pa @1KHz (-74 +/- 3dB; 0dB = 1V/0.1Pa).

The output is balanced, i.e., neither side of the coil was connected to the case. Dynamic microphones are a mature design and have been around for decades. Unlike the common and inexpensive electret mic element, the dynamic element does not require external power to operate. It generates its own electrical signal. One can consider they operate backwards to a speaker.

Dynamic elements are reasonably rugged and can be inexpensive. They are used in a wide variety of consumer, industrial, and professional devices from junk to extremely high quality. They are easy to work with and can be excellent performers if incorporated into a properly engineered system.

Comments—The size of the device/microphone element indicates it was intended to be a part of a relatively small device. Obvious high-quality construction indicates the device was intended for professional use. The alphanumeric markings are characteristic of high-quality devices where tracking part numbers, references to engineering drawings, perhaps serial number or manufacturing lot number, specifications, etc., must be known. I do not recognize the markings. Appropriate investigation undoubtedly could uncover the exact manufacturer of the element, and the device(s) for which it was designed. This investigation is beyond the scope of this report; however, I would be glad to discuss further effort if you so desire.

I would expect this element to be manufactured by one of the professional microphone manufacturers such as Telex, Electrovoice, Shure, Astatic, Audio-Technica, or the like. I have absolutely no indication of who the manufacturer may be at this time; I merely mention a small number of companies who work with precisely this sort of product.

Considering all factors mentioned, my supposition would be the element was intended for use in a wireless mic, of the sort used by active speakers, preachers, etc., whose movements might be hampered by a wired microphone. I believe, but do not have any direct evidence, that the microphone would have been intended for an entertainment or sound reinforcement application as opposed to communications, consumer, etc., use.

Please note this element by itself is not a self-contained eavesdropping device. It could be a component of one. Two conductors would have to connect the microphone element to an amplifier, transmitter, recorder, etc., to process the signal generated by the microphone.

Please note the comments following are mere supposition or my opinion only, not based on any direct knowledge of the instant situation but merely on my experience and training.

The mic element was discovered magnetically attached to a metallic serving tray of 31.5 square inches (800 mm). This is a significantly larger area than the tiny membrane on the element itself. Standard geometry indicates an approximate area of the microphone's pickup element of 0.125 inches (3.17 mm). Therefore the area of the metal tray is approximately 250 times of the area of the element's diaphragm.

Magnetically attaching the element to the serving tray would greatly increase the apparent sensitivity of the microphone element by creating a much larger area to capture sound pressure. I was able to demonstrate this in testing. A very noticeable increase was obvious. The membrane/diaphragm side of the element would stick to the tray fairly intimately.

However, all this still leaves the factor of needing to connect to the output of the mic element in order to make any use of the sounds picked up by the element and converted to electrical energy. No indication of soldering or scratching from mechanical clips to the element's spring-loaded terminals is evident. Therefore, one would presume any connection would be as originally intended: by spring contact, clamping, or some other mechanical means. While this could be possible in theory, it seems rather unlikely in an uncontrolled situation such as casual placement in a conference room. However, the officer assigned to the investigation should make a thorough inspection of the area in which the mic element was discovered, looking for any signs that could indicate that some sort of spring/friction connection could have been possible. Again, more in theory although absolutely possible, strips of conductive paint on a flat surface could have carried the mic's signal to a location where the information could be processed. A second theory is the element merely was someone's toy obtained from disassembling some device in which it was contained, and simply lost.

Another theory could be the element was purchased from unscrupulous sellers purporting it to be a self-contained surveillance device and installed under that presumption. Considering the generally sleazy nature of persons selling illicit surveillance devices to the public as well as the usual lack of technical sophistication of the buyers, this is not an unusual scenario. The buyer, if they would discover the unit is not a surveillance device as claimed, certainly cannot take legal action against the seller. A simple review of spy shop Web sites will reveal hundreds of useless devices advertised to do all sorts of magical things a layman might believe but which would be obvious to a professional as vaporware or hyperbole.

Yet another theory is the device could have been left by persons presently unknown to be discovered by company personnel as a terrorism tactic. Thus far, the device has accomplished precisely that purpose. Numerous high-level people have been involved, meetings have been held, and I was consulted on the case. All this results in a certain amount of disruption, expense, and concern to the victims. Therefore, even though the instant device is worthless on its own, it already may have accomplished its mission.

A final theory, and one which may deserve the most consideration, is the device is a "throwdown" intended to be discovered while a genuine, functioning, effective surveillance device is in place operating as intended. It is fairly common for one or more easily found devices to be left in fairly obvious places, with the hope an inexperienced security or countersurveillance practitioner will find it or them and consider them to be the only devices installed. The genuine devices continue to function.

As I am not an officer assigned to this investigation, I have not had the opportunity to inspect physically or electronically the area in which the device was found. Therefore, I cannot comment further. However, the government officer who contracted me to examine this device has some formal training in TSCM and is fairly well read but admittedly has limited actual field experience. His report and opinions should be considered along with my comments here, and the composite used to develop a profile of the device and circumstances surrounding its discovery.

I recommend, if the information contained in this report is considered to be worth pursuing, a competent TSCM sweep of the facility be considered. Whether in current use or not, a sweep may reveal additional devices, genuine surveillance devices, or evidence of such having been installed.

I would be glad to recommend a competent TSCM practitioner for this work if the decision is made to pursue that course of action.

Please contact me if you have any further questions in this matter.

Together with this report, I am returning the microphone element and tray to you via traceable courier along with an invoice for my services. Thank you for allowing me to assist in this interesting assignment.

Respectfully submitted,
Steve Uhrig
for SWS Security
enc: as stated

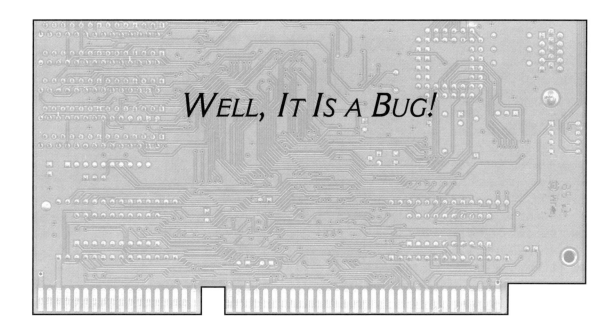

Well, It Is a Bug!

Not long after one of their employees had left the company, the owners of an escort service became aware that something strange was happening.

The dispatcher would send one of their girls in a limo to meet the customer but when she arrived, the customer was gone. This former employee (she called herself "Tiffany," by the way) would pretend to be from the escort service and would take the customer for herself.

The owners started to catch on and wonder if maybe someone was somehow monitoring their phones to steal their business. They decided to call in Rick Hofmann to sweep their premises. "My client was losing both money and reputation," Hofmann said.

It didn't take long for Hofmann's professional TSCM team to find the problem, and it wasn't a wiretap; it was a bug. Actually, it was two bugs—an RF surveillance transmitter built into a clock and a transmitter hidden in a coffeemaker. With these bugs, Tiffany was able to overhear the dispatchers and beat the limo to the pickup location.

The desk where the clock was located belonged to the dispatcher, which meant the spy could hear details of client meetings, and the coffeemaker was a place where office personnel congregate, so the spy was able to stay on top of what the employees were talking about. The office gossip was, naturally, about how they were losing business.

The coffeemaker was between the two dispatchers' desks, which allowed it to pick up audio from both dispatchers. The clock was on a desk behind the dispatch desks.

It was never determined where the listening post was, but given the details, it was most likely a car or van parked outside the escort service offices. It could have been in another office that was within range of the transmitters, but there was no way to know without a complicated and expensive investigation and a way to set up and catch the bugger.

Now, if you have read any of my other books on surveillance, you know that sometimes I will describe a situation and give you part of what I know—what I have learned from this particular experience—but not everything. At least not right away. The reason is that I am encouraging you to try and figure things out for yourself—to analyze the situation and apply what you have already learned and find the rest of the answers.

It was a fairly sophisticated black-bag job. The person who set it up had a working knowledge of basic electronics and had probably researched the operation and determined that the transmitters he used would have sufficient power to reach the listening post. Probably. There is always the possibility that he or she took a chance and purchased what was available, hoping that it would work. We will never know. What counts is that it did work, and that is the objective.

The design of this transmitter indicates it was mass-produced.

OK, first, if you look closely at the picture of the transmitter, you will see that it has surface-mount components. It costs a great deal of money to design and manufacture a surface-mount device. This means that it was not a one-time thing, it was not homemade; it was mass-produced.

Again, look closely and you see several crystals. This indicates that the transmitter was capable of sending its signal on more than one frequency. Why would the spy use something with more than one channel?

Again, maybe because it was the only thing available. But on the other hand, to an experienced countermeasures technician, this indicates that there may be other similar devices. If there are two bugs operating on the same exact frequency, one will interfere with the other. Something called the FM "capture effect." So, the spy would need either two receivers or a scanner programmed to alternate between the two.

Now, consider that according to statistics from *Wiretap Report* on how many legal surveillance operations there were, that RF

Even before you see the inside workings, something's not quite right about this clock.

transmitters were used in only a few of them. So, it isn't logical that whoever spied upon the escort service was a federal agent—the feds wouldn't likely spend the money to have mass-produced bugs. They would build their own or get them from Innovative Surveillance Technologies or DTC, Tactical Technologies.

So what are they already? The transmitters are wireless microphones sold by a well-known nationwide chain of electronics stores.

There is one other thing to consider. Think about what you know about bugs and look at the picture of the clock. What doesn't seem right, what may or may not be right, and why?

The answer is somewhere in another chapter, cleverly hidden away.

Part Four

THE INTERNET

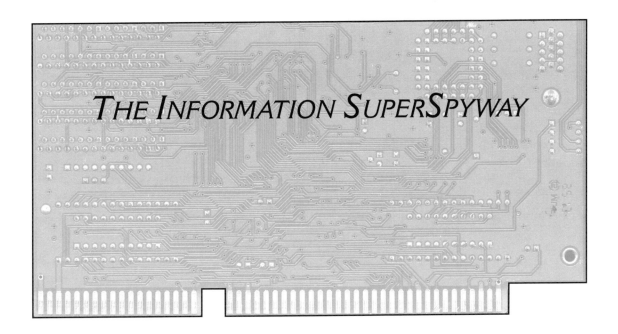

The Information SuperSpyway

BACK IN 1993 I WROTE A BOOK CALLED *THE PAPER TRAIL*.[1]

Not an original title, but accurate as it contained a great deal of information on how Big Brother and Little Brother, corporations large and small, amass details of people's personal lives.

It was also about how you can find out, to some extent, what they know about you, how to get information on other people, and mostly, how to drop out; how to stop leaving an information trail behind you, how to disappear; to use some unconventional methods of covering your tracks, and to take it on the lam in such a way that you would be very difficult to find, even by law enforcement agencies. After all, while the FBI does have a Ten Most Wanted list, they don't always "get their man."

The Paper Trail was written around the time that the general public first learned that the Internet, which had been a well-kept secret for decades, existed. So, it seems that some politicians decided that there should exist an Information Superhighway. It was described as a system through which students would be able to access sources of information not previously available to them, that physicians in remote locations could use to send medical information to major health care facilities for diagnosis, and that people everywhere could use for a host of other benefits ...

And the Internet became public.

Now virtually everyone in the United States knows that the Internet exists, and the majority of people have access to it, one way or another.

I knew when I wrote *The Paper Trail* that the Internet would soon become the greatest source of personal information about people that ever existed—the most effective tool in the government's arsenal of weapons in the war on privacy. But back then, the Internet wasn't used to gather personal information on individuals except those who chose to make it available. There was, and is, something called Finger that anyone who wanted could use to leave a short bio of themselves, a resume, contact information, whatever they wanted, that was available to anyone who used the Finger tool. More on this coming up in the chapter "Internet Tools."

Now, in 2005, many people are starting to realize just how true this—unfortunately—is. But there are also many who do not and are unaware of just how much information can be obtained about them from the Web sites they visit, the questionnaires they answer, forms they fill out, and seemingly innocent little banner ads that capture their curiosity.

Make no mistake about it. Everything you do on the Internet is being monitored. Records are being kept. A profile built.

Remember—when you hit the Enter key, you have no control over where your information ends up. But there is a great deal you can do to disguise it and yourself using simple tools, many of which are available free.

Read on and learn about making the Internet a safer place to be.

ENDNOTE

1. *The Paper Trail*. M. L. Shannon. Lysias Press. 1993. ISBN 1-884451-09-3 (Out of print.)

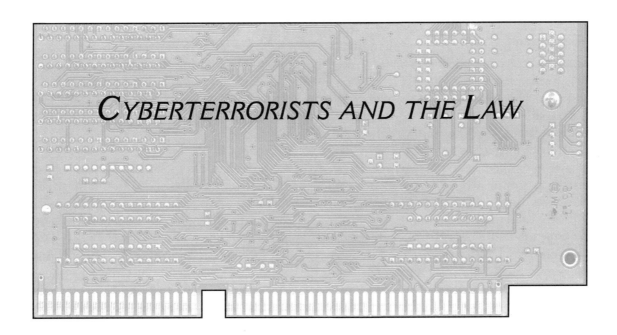

Cyberterrorists and the Law

People rob banks.

It doesn't happen very often; you probably have never actually seen someone come running out with a gun in one hand and a sack of money in the other, but it happens. And the government wants to stop bank robbery completely, right?

Armed guards wandering about the lobby won't stop someone from sticking a gun in some poor terrified teller's face. So, they install metal detectors at the doors. Fine—that keeps people from bringing guns into banks.

Anyone remember the famous zucchini bandit? He stuck up a bank by stating that he had a gun; he got caught and what he had in his jacket pocket was, right, a zucchini. So, one of these days, if it happens again, maybe some politician will decide to spend millions of the taxpayers' money to build a zucchini detector.

Word gets around, so the next guy uses a cucumber.

You could strip search everyone and that might convince lawmakers that they have the answer, although this would make the customers a little less than thrilled. And it wouldn't work. Harry Houdini mastered something called "controlled regurgitation." He could swallow a lock pick and then cough it up to open the handcuffs or whatever he got himself into and out of. Suppose someone learns this technique, walks into First National, barfs up several glass tubes of poison gas, and demands the cash?

Completely stopping bank robbery is like ending so called cyberterrorism—just can't be done. But with the government's mentality they have begun a sort of cyber-strip search. They have taken away our privacy, our Internet freedom, and we have barely even seen the beginning. In England there was a plan to archive every e-mail sent through the Internet—at least that goes into or out of the United Kingdom. I suspect a similar plan is already in the works here in the United States, meaning that you will never again be able to send private e-mail unless you encrypt it. And, just as they have for decades, the government is trying to take secure encryption away from us.

Making more laws is not the answer. More raids like Operation Sundevil are not the answer.[1] Taking people's computers and throwing them in jail—even if there is no evidence that they committed a crime or that a crime has even been committed—is not the answer.

So what, you ask, is the answer?

I read somewhere that the feds are planning a closed system. Something that would link government offices, facilities, airport air traffic control, electric distribution installations, nuclear power plants, and other facilities likely to be targets of terrorists. It would be a secure network that would have very limited access. And guess what? This is exactly what the Internet started off as, the Department of Defense Advance Research Products Network—DARPAnet. A closed network using fiber-optic lines, ultrasecure encryption, and tightly controlled access that would take the heat off people who might otherwise be wrongfully prosecuted as malicious hackers.

Strong encryption in data sent from one government office to another, even over wireless networks using something like DESx or PGP, if used right, is unbreakable by brute-force attack. So why aren't federal agencies using them?

One problem could be the lack of employee awareness about the potential danger of using unsecured channels for sensitive material.

And, of course, government agents seem to lose laptop computers frequently.

Secure encryption methods could go a long way to prevent terrorists from obtaining information they are not allowed to have.

Terrifying some teenagers by kicking the door in and threatening them and their families with machine guns is not the answer.

ENDNOTE

1. In March 1990, the U.S. Secret Service raided a number of homes and businesses, including Steve Jackson Games in Austin, Texas, as part of a nationwide investigation on data piracy. Many computers and disks were taken and several arrests made. The Secret Service claimed the "criminals" were guilty of numerous offenses, but later it was revealed that, in fact, no laws had been broken. This case resulted in the creation of the Electronic Freedom Foundation, and more information can be found on their Web site at http://www.eff.org/. In February 1989, Craig Neidorf, a student and publisher of an e-magazine called *Phrack*, was prosecuted for revealing "confidential" information about the 911 emergency phone system. The government continued to persecute Neidorf even after it became known that this information had nothing to do with hacking into the 911 system and was neither confidential nor technical; it was strictly administrative and had already been published in a $7 book available at, among other places, the Stanford University bookstore.

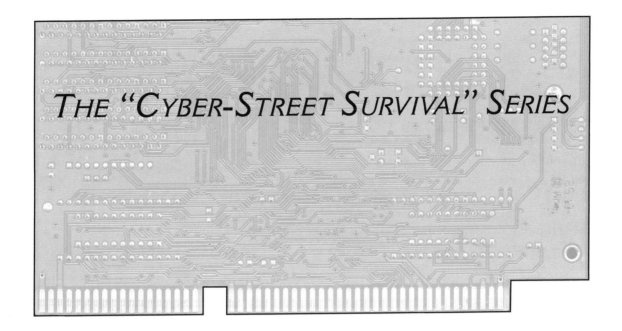

The "Cyber-Street Survival" Series

THIS CHAPTER OF THE BOOK IS BASED IN PART ON A SERIES OF ARTICLES I wrote for *Nuts & Volts* magazine, which were published in the January through June 2001 issues. Reprints are available from the publisher by visiting http://www.nutsvolts.com. Here is a rehash of the series and what was covered in each article, along with some recent comments.

PART 1: GETTING STARTED

This article includes an introduction to the Internet; how the 'Net came to exist; how to get online for the first time; basic information on privacy; and why the Internet is insecure.

Since the original series, privacy on the Internet has virtually ceased to exist. I repeat: When you hit the "send" button, you have no control over where your information goes or where it ends up.

PART 2: SPAM: JUST SAY DELETE

This part was devoted almost entirely to spam and ways to deal with it.

Not much has changed since the original series, other than that spam has increased by a few orders of magnitude. Yes, there is this new tool and that new tool that you can buy that promises it will eliminate spam, and while some of them work, there is still the problem that they will block legitimate messages.

Some providers have set up systems to mark and/or block spam; the filters used in Yahoo! mail are excellent, but they, too, may very well block your important e-mail.

And while it is a hassle sorting through spam to find the legit e-mail, it is better than losing something important, like a job offer.

So, "just say delete."

PART 3: WITHOUT A TRACE

With your home computer, using dialup or DSL, you leave a record of your IP in the log files of every site you visit, as well as at the ISP where you have your account and certain government agencies (Homeland Security and all that). By using proxy servers for Web surfing, you could connect through or be "relayed" by a true anonymous proxy without leaving an electron trail.

In a wireless world, this has become much easier, what with so many Internet cafés where your electron trail leads back to the place you are having lunch, but not directly to you. If you know how it is done. See the chapters on wireless networking.

PART 4: SECURITY AND OTHER THINGS

More on security; how to use a packet sniffer to see what is coming into and leaving your computer; general information on cyberstalking and prevention; and a look at those who would force their beliefs upon the rest of us by passing laws restricting what we may or may not see on the Internet.

In this book, the same rules apply, but the emphasis is, of course, on Wi-Fi, or wireless networking.

PART 5: HACKERS

A journey into the mystical world of hackers; why you should be damn glad there are hackers; plus media myths, distortions, and lies; encryption and making your files and e-mail secure from intruders; Trojans and preventing them; and the basics of data encryption.

In this book you can read about someone who attacked my wireless network as well as how I was able to take control of another network.

PART 6: INTERNET TOOLS

Part 6 is about programs that you can download and use to learn more about the Internet, find out who is behind a Web site, and maybe trace e-mail to its source. Here, you can apply them to wireless and learn about some excellent applications for gaining a solid understanding of the wireless world.

The entire chapter is reproduced beginning on page 91.

Yes, the original series was written for beginners or those with very little knowledge of the Internet, and before the wireless world revolution. This new series is likewise intended for beginners, but it goes well beyond the very basics. It won't make you an expert. It can give you a good working knowledge of these subjects.

I am working on several new articles for the Cyber-Street series and will submit them to *Nuts & Volts* soon.

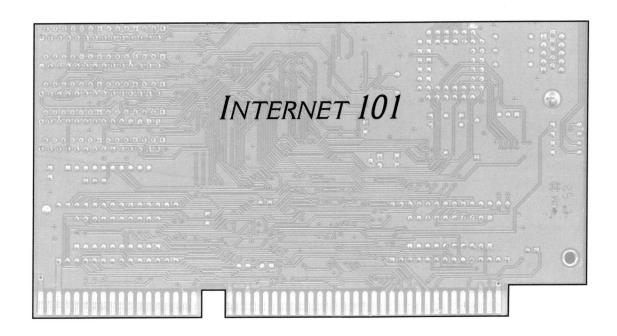

Internet 101

More and more people have Internet access every day, and some of them are finally beginning to take the measures needed to walk on the safe side of Cyber Street. Some, but by no means all.

So read on, even if you are experienced on the Internet, even if you feel secure that no one is likely to hack his way into your computer. You just might learn a few things.

In the "Cyber-Street Survival" series I explained how any time you connect to the Internet there is the chance that someone is keeping track of your movements through the World Wide Web or FTP or Telnet or any other TCP protocol you use.

Someone may be sending you files that end up being stored on your hard disk without your knowledge or analyzing your e-mail looking for certain key words.

Every time you log on, you may unknowingly be inviting spammers to bombard you with phony ads, get-rich-quick schemes, chain letters, and invitations for you and your children to visit pornography sites. Each day as you read your e-mail, you take the risk that a message may contain a virus that could wipe out every file you have. Whenever you are online, there is the possibility that someone will be scanning you, looking for open ports—ways to invade your computer—and trying to install a Trojan horse with which they will be able to take control of your system. Perhaps to see what's there, make copies of your files, or destroy information.

And most insidious of all, there are Internet stalkers out there on Cyber Street who may be able to harass you, even steal your identity and impersonate you, which could have tragic consequences. Most data can be restored, but does your reputation and credit rating have a backup file?

Why do these things happen? Because people, without knowing, make it easy for hackers, who know more about the technical details of data communications than does the "average" user.

And though the number of people on the Internet is increasing at a phenomenal rate, in spite of the warnings on the new media, most of these users still know little about the 'Net or the computers they use. And many of them are unwilling to take a few hours to make their machines secure against these attacks and abuses. Too lazy.

Too busy. Or they simply do not know how. And the heck of it is, too often they don't learn from a bad experience.

Most anyone who has ever had his or her home burglarized will take preventive measures to keep it from happening again. But this doesn't seem to apply to the Internet. Remember the "Melissa" virus? And then along came "The Love Bug" and the same thing happened. People, thousands of them, had their computers infected. And it will happen again.

> AMONG THE VARIOUS THINGS THAT VIRUSES can do, the worst is to format your hard drive, destroying everything that is on it. To do this, it uses a program on your computer, called format.com, com being a command file.
>
> Format.com is somewhere within Windows, depending on which version you use. Win 2000 has it in the WINNT/SYSTEM32 directory.
>
> If you rename it to something like format.xxx, then it will not execute; it will not run.
>
> An excellent application for file management, including renaming files, is Ztree for Windows available at http://www.ztree.com. If you try it, like millions of other users you'll wonder how you ever got by without it. It really is that good!

Some people are finally learning to lock their digital doors, but others go crying to the government to pass more laws to protect them.[1] Which, they don't realize, is the antithesis of what the Internet is all about. Or was. Once, the Internet was a self-regulating entity. But now, because of all the abuse and the problems, government and big business are passing laws that say what we can and cannot do. This is even more so since 9/11, TIA, Homeland Security, and whatever else the government is up to in order to protect us from terrorists as well as from ourselves.

With the information here, you will be able to keep most of these things from happening to you. You will be able to surf the Web anonymously so no one will know where you are coming from or where you have been; encrypt your e-mail so that no one but the recipient can read it; make your computer virtually invisible to potential hackers; keep your personal information private; deal with unscrupulous marketing companies; and reduce the amount of junk electronic mail (spam) you get.

You can also learn about some programs with which to trace spammers, find out who owns a Web site and where they are, who is behind a banner ad, and other interesting things. You can install a "packet sniffer" program that displays information entering or leaving your computer that you otherwise would not be aware of, and also take a brief excursion into the murky and media-distorted world of hackers.

No technical knowledge of either the 'Net or computers is necessary to take advantage of what is here. Other than an Internet connection, you need only to be able to download and install software. The programs that are reviewed here will have step-by-step instructions, and there are links to many sources of FAQs and Help files.

GETTING STARTED

You have Internet access from your own computer. And regardless if it is dial-up, DSL/cable, or wireless, you can do these exercises. (Note that setting up and securing a wireless network is in a separate chapter, so I may repeat a few things here)

First, let's see how vulnerable your computer is. Go to Gibson Research Corporation at http://grc.com. Scroll down to the dialog box that says Shields UP!! and follow the instructions. You can check all of the service ports—the ones used most often for Internet data. It will take a few minutes, then you will see what ports are open to hackers to attack; there will probably be quite a few. Read what it says about Net Bios and Port 139, which are especially important.

FIREWALLS

Next, go to http://www.zonelabs.com and get ZoneAlarm.[2] This program is a firewall that monitors attempts to connect to your computer and pops up a warning message. Download it to a directory of your choice, then use File and Run from Desktop and it will install itself. Restart your computer and it will load automatically. You will see a square icon, red and green, on the task bar at the bottom right corner of your screen. There are settings you can change and which are explained in the Help file or at the Zone Labs site, which you can experiment with later, but this is not required. It is already running and protecting your computer.

Now, go back to Gibson Research and run the tests again. You'll see that the ports tested are now closed or even in stealth mode so that when someone scans your IP address, they won't find anything—no indication that these ports even exist. But for now, it is important to close the ports that Gibson may have found to be open.

Click on Start, then Settings and Control Panel. In Control Panel, double click on Network. You will see a dialog box and near the bottom, a smaller box that says:

File and Printer Sharing and then:

- I want to be able to give others access to my files
- I want to be able to allow others to print to my printer(s)

Both of these selections have to be unchecked. Unless your computer is part of a network, directly connected to other computers, there is no reason for either of these to be enabled. They probably are checked by default, and this is one of the most dangerous port conditions you can have.

Once this is done, restart your computer and reconnect to the Internet. Disable ZoneAlarm and go back to Gibson Research to repeat what you just did. You will see that the Net Bios port is now closed. You have just made your computer very secure against attack. More so than millions of other computers whose owners have done nothing. Later in this series, we will go into other types of attack using other ports, as well as attacks through e-mail. But for now, here are a few more important things you can do:

If you receive any kind of an attachment to an e-mail message and you don't know for sure whom it is from, don't open it. Copy it to a floppy disk for future reference if you like, and then delete it.

If you are using Microsoft Outlook, be aware that it is more vulnerable to viruses than most programs because it is the most used and the most attacked. Download and install an e-mail program such as Eudora or Pegasus, which is free and available at http://www.tucows.com. Both are less likely to fall victim and, just as important, less likely to pass a virus on to those in your address book.

A good place to start learning is the Internet Explorer Security Center at http://www.nwnetworks.com/iesecurity.htm. Some of the default settings are the reason that Melissa and The Love Bug caused so much misery. Better yet, don't use Outlook or Internet Explorer at all.

Then read the Help files in your browser. Learn about ActiveX and Java Script. Use the configuration files and make changes so that your real name and e-mail address no longer exist. Web browsers can read things that are hidden inside e-mail that is written in HTML (HyperText Markup Language, the language that browsers understand).

Internet Explorer is also a target for hackers and is more vulnerable to attack than other browsers. Consider Firefox, which is a free program also available at Tucows.

Now, let's have a look at some privacy issues and how to handle them.

LOG FILES

Whenever you connect to a Web site, you leave behind a certain amount of information about yourself. This may contain any or all of the following:

- Your IP address
- The date and time you logged on and how long you stayed; the length of your visit

> SENDING YOUR SOCIAL SECURITY NUMBER to anyone, for any reason, on the Internet is a very bad idea. You are asking for trouble if you do. A little research with Google will bring up many horror stories from people who made this mistake.
>
> If you want to use online banking, please consider setting it up by telephone or in person so that you don't need to send the entire SSN over the Internet—only the first five digits.

- What areas of the site you viewed and the sites you were previously visiting
- The type of computer you are using and the operating system (Windows, Mac, Linux)
- The Internet Service Provider where your account is and the city it is located in
- Any words you used to make a search to find the site
- And, depending on how you set up your browser, even your real name and e-mail address

FORMS

Every time you fill out a form at a Web site, you give away information about yourself and perhaps your family. Now, you may see something called a privacy statement in which you are assured that your information "will not be sold." Perhaps not. But it may be "given away," in exchange for similar information from another marketing company. Traded, in other words. Or loaned. Or rented.

BANNER ADS

Banner ads are on practically every Web site you visit. And they are not as harmless as they may seem. Clicking on them often provides information about yourself.

"How many times has your heart beaten in your lifetime?" You are curious so you type in your day, month, and year of birth. Your Web browser may have already told the company that placed this banner your name and e-mail address, and now they have your birth date. Meanwhile, a cookie (explained later) has been placed on your hard disk drive, which you may not have known about, and the next time you log onto that site, you see banner ads oriented toward people of your age group.

> AN EXERCISE:
> Spend a few hours roaming around, checking out various sites and look at the banner ads you find. Then think about what details of your life might be revealed if you answered their questions. And don't believe for a minute what you are promised about privacy and confidentiality.

OK, big deal, you say. So what? Another banner ad implies that the company behind it is concerned about your health. They have a nifty little chart where you enter your height and weight to see if you are within certain standards. You'll probably lie by at least a few pounds, but still, this is one more thing "they" know about you.

Another banner wants to know what kind of car you are thinking of purchasing. Now they have a good idea of your income bracket.

Yet another banner ad expresses concern for your health, so they make it easy to "find a doctor near you." To provide the location, they want your Zip code. Marketing companies are really big on Zip codes. They have sophisticated databases and charts and diagrams that tell them how many people live in a given Zip, the average income, type of housing (homes or apartments), real estate values, occupations, vehicles owned by make and value ... And don't forget, by knowing what kind of doctor you require, they now know something about your medical history.

Banner ads offer to get you "guaranteed low interest loans," but you may have to type in your Social Security number. Now they know virtually everything about you—since technically you have applied for credit, they can access part of your credit history file called a "Header."

AD BLOCKERS AND OTHER PRIVACY APPLICATIONS

A number of programs have been produced that help prevent pop-up banner ads, defeat Trojans, etc. AdAware (http://www.lavasoftusa.com/), which detects spyware, and The Cleaner (http://www.moosoft.com/products/cleaner/), which defeats Trojans, are both good.

Another great little program is Zeroclick, available at http://www.privsoft.com. It blocks banner ads and cookies from DoubleClick, one of the big names in these nuisance ads. Zeroclick promises that "...You'll never see their banner ads, you'll never get their cookies, and your machine cannot send DoubleClick any information. DoubleClick ceases to exist..."

COOKIES

A cookie is a small file that contains information about you and, unless you block it, it is placed on your hard disk drive. Now, this information may be used only at the site that placed it there. You visit that site again, their computer reads the cookie and has an idea of your interests, what areas of the site you visited before. And, using that, it can take you directly to those areas. In other words, all cookies are not necessarily "bad"; they may be a convenience to their visitors.

However, "third-party cookies" don't stay at that site. They are sent to marketing companies that have deals with thousands of sites to forward this cookie information to them.

Log on to a Web site and watch the status window at the bottom of the browser screen. As it is loading, you will often see the names of well known companies flash by: DoubleClick, of course, also Flycast and GotoNet.

Jason Catlett, president of Junkbusters Corp., said, "Cookie leaks are the bug from spammers that keeps on bugging. It's intolerable that e-mail can be used to silently zap a nametag onto you that might be scanned by a site you visit later. It's like secretly barcoding people with invisible ink."

So, again, if you are using Netscape, create a new profile. Don't use your real name. Call yourself Sally Sikorsky or Joe Jacovitz or whatever. For an address, just leave it blank or enter something like joe43@flibberdygibbit.com. In other words, use an address that doesn't exist. As to the old profile, you can't delete it and reinstalling Netscape won't get rid of it, so just ignore it.

While you are at it, set Netscape so that it does not accept cookies. OK, some sites will refuse to let you connect, so unless it is really important that you do, then just leave. Whatever you can find at one site, you can probably find at another. If you have no choice but to accept the cookie, then delete it after you move to a different site. If you are using the Opera Web browser, you have the option of selectively blocking third-party cookies that go somewhere other than the site you are presently visiting. Opera also offers the ability to delete cookies when you close it or disconnect from the Internet. You can get Opera at http://www.tucows.com; it has a free trial period and registration is $40.

DOG COOKIES

Scotty loves cookies. He loves to eat them.

Scotty is a little black dog who sits faithfully on the taskbar at the bottom of your monitor screen and sniffs at things periodically. If a new cookie is received, "Arf-Arf!" Scotty pops up to warn you. Scotty is part of a really neat little program called WinPatrol available from http://www.winpatrol.com.

These little things that take a minimum of time to implement and cost very little will make a big difference in the time you spend on Cyber Street.

PROXY SERVERS

When you connect to the Internet with whatever browser you are using (hopefully Opera or Mozilla FireFox and not Microsoft Internet Explorer), as you read above, you leave a record. Log files, including the IP address of your computer.

Now, suppose you don't want to leave this data trail. You want to surf (I don't know where that term originated) anonymously. To do so, you can use a proxy application, which is a program you install on your computer.

What this does is, instead of connecting directly to the Web site you want to visit, it forwards your request to any of hundreds of proxy servers and, from there, you connect to the site you want to visit. And in the log files of that Web site, the IP that is recorded is that of the proxy, which could be down the Cyber Street a few blocks or could be thousands of miles away.

One of the best applications is called Anonymity 4 Proxy (A4Proxy), which is available at http://www.inetprivacy.com.

It is an excellent program, one of the best and most affordable. I have used it, in its various versions, for years whenever I didn't want to leave a trail behind me in researching controversial subjects.

A full-featured trial version is available for download, and registering costs only $45, with which you get a massive database of anonymous proxy servers as well as excellent tech support.

I highly recommend A4Proxy. So, if you want to cover your tracks, for whatever reason, I hope you will try and buy.

OK, having read all this, and hopefully done some serious reading, now is a good time to go back to Gibson Research and try some of their other security applications. They are free and they are damn good, but—and make a note of this—make sure you read through the entire Gibson Research site, all of the articles and Help files; everything there. It will take time, as there is a lot to learn, but, and I again stress this, it is important that you do so before you use the applications. If you do not, it is possible that you might lock up your computer—or even your whole network—from making an Internet connection. Nothing that you do seems to fix the problem, and unless you have another computer that does have an Internet connection, you might be resigned to re-installing your operating system.

Please: RTFM!

Later, in the chapters on wireless networking, I will talk about routers, some of which include a hardware firewall. Even if you have only one Internet computer with a DSL connection, and so don't actually need a router, please consider purchasing one. They don't cost much (some are available for $100 or less) and they provide an added measure of security.

> YEAR AFTER YEAR WE READ about the security holes in Microsoft Internet Explorer and how tons of viruses target Outlook Express. What does it take for people to finally realize that if they stop using these applications, they will be far more secure from viruses and other exploits?
> When will they ever learn?

Now that you know how to secure your computers, let's take a look at identity theft.

ENDNOTES

1. http://www.fusionsites.com/written/Humor/wood/wood.html

2. About ZoneAlarm: I have used it for many years and have depended on it to block hack attacks, which it has done. I have even invited hackers that I know personally to try to get into my home network. No one was able to do so.

Just before this book went to the publisher, I saw some reviews of the latest version of ZoneAlarm that were not encouraging. According to these reports, it may no longer be as secure as the earlier versions. And because of my deadline, I have not been able to test the new version. So, may I suggest that you learn all you can about the new ZoneAlarm before you lay the security of your computers on the line.

Nor was there time to review any of the many other firewalls available.

I did download PC Viper from http://www.sourcevelocity.com, which looked promising, but I couldn't get it to install, and e-mail to their tech support was never answered.

ConSeal is another excellent firewall; it was the very first one I used, even before ZoneAlarm, until it was sold to another company. (Who, in my opinion, really mucked it up.)

Well, it is back, under the name 8Signs Firewall available at http://www.consealfirewall.com. Now be aware that while it is a very powerful and versatile firewall, one of the best, it has a learning curve. So expect to spend some time getting it set up and running with all its bells and whistles. Just as I plan to do now that this book is complete. And it just might replace ZoneAlarm on my network.

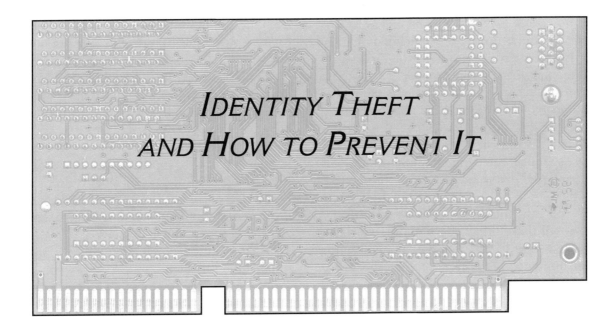

Identity Theft and How to Prevent It

There was a story on one of the Internet news services that told how a computer programmer went to an automobile dealer to buy a new car. So while the credit manager checked up on him, it was discovered that someone else was using his Social Security number (SSN)—that "someone" being a suspected terrorist. And the programmer was in a world of hurt trying to get it straightened out.

This "someone" may have just made up a number at random and it happened to be issued to the programmer, or may have seen the guy's number somewhere. Perhaps on the Internet.

This will probably have a happy ending when the guy gets it straightened out, but other stories end in tragedy. And as you can see in the media, they aren't that uncommon.

There are numerous ways people can get information about others—such as you and members of your family—in order to "become" them. Use their name, the many numbers that are attached to people in the form of little plastic or paper cards. Digging through your trash, called Dumpster diving, is one way, so be careful what you throw away. But for some hackers the Internet is a virtual Dumpster because of all the information that is stored there. Such as the airline company that made a security mistake causing thousands of customer records to be available—even to those with no hacking skills. It happens and it happens often, but just as often there is a cover-up and the public never learns about them. They just hear about the people whose information was snagged by someone looking for new victims to exploit. And remember about bank robbers—there is no way to make the entire Internet secure against these incidents.

Yet millions of people place personal info, driver's license, bank account, SSN, and other numbers on Web sites because their credit card numbers are required to sell on online auctions and their home addresses are required in order to have online purchasers shipped to them.

So what you need to ask yourself is that even though the odds are way against having this happen to you, if you refuse to put your numbers online, the odds become near zero. Even if someone makes up an SSN that is yours, unless that person is being watched by law enforcement or otherwise is something other than "just plain people," the odds are nothing will come of it. At least nothing serious. A woman in Indiana was using my SSN for a time but it didn't cause me any problems. And maybe she even paid some money into my account!

No one can get something from the Internet if it isn't there. And that is the first step in preventing identity theft.

OK, you say, fine, but it's too late, I already have done all those things. So undo them. Have your bank accounts closed and re-opened with different numbers. Ditto credit cards. The extra time and trouble it takes to mail checks instead of paying online is well worth the peace of mind you will have.

If you absolutely have to make online purchases or require a credit card for auctions, consider this: Open a new bank account and get a debit card. Keep only as much in that account as you will need for specific things—you can make transfers by phone, you know—so that if someone gets the number, it isn't of much value to them. Rent a post office box and have all mail concerning the Internet and your banking affairs sent there along with your online purchases.

Again: No one can get something from the Internet if it isn't there.

What Is Hacking? What Is a Hacker?

There are many definitions, many variations on the theme, depending on whom you ask. A hacker is someone who likes taking things apart and putting them back together to learn about them, to modify and improve them, to experiment. Hardware, for example. This writer can be considered a hardware hacker; the cover of my main machine, a Pentium II, has never been installed. It sits there on the floor with several drives lying on the bottom or hanging from the bays, with wires and cables hanging outside the frame, and various odds and ends of radios and test equipment connected to it.

A programmer (which I am not) is a software hacker. Someone who does the same things with code as a hardware hacker does with circuit boards and drives and cables.

If this definition serves to cause the reader to believe that not all hackers are malicious, destructive, or dangerous, then the reader may gain a great deal from this book. If not, then perhaps the reader should donate the book to someone who will profit from it.

The real definition of a hacker is an attitude, a mind-set, a way of life. Hackers tend to dislike authority and rules; to be unconventional and often clever, if not inventive; to know things we are not supposed to know and get into things we aren't supposed to; to be places we are not allowed to be.

A hacker is someone who takes things apart to see what's inside and how they work and then (usually) puts them back together. In other words, the term applies not just to computer geeks but to people like Bell, Edison, DeForest, Tesla, and Farnsworth.

With a few exceptions (and they are indeed few) we hackers are not malicious. We are not destructive. We do not sit gleefully in front of our equipment ingesting massive quantities of Jolt cola and chain smoking Benson & Hedges, plotting against the Cyberworld. Finding things to steal or destroy.

But ... I heard on TV ... But "they" said on the radio ... But I read in the newspaper ...

The mainstream media is not in business to bring you the news; they are in business to make money. The same as any other business. So, in order to sell papers, magazines, in order to get people to watch the news on TV, they have to be able to report things that get people's attention. The more tragic, the more sensationalist the stories, the more of their advertisers' products they sell. And as people are starting to realize, thanks to the Internet, there is often a great difference between what is reported and what really happened. And it is not unusual that some of the stories broadcast are total fabrications.

"Hackers take control of a British military communications satellite."

No. Didn't happen.

"Teenage hacker holds U.S. corporations hostage. Demands and gets Ferrari."

Uh-uh. Nope. Didn't happen.

Kevin Mitnick was variously described as evil and dangerous and whatever else. He was accused of stealing software that cost several big corporations hundreds of millions of dollars. Funny thing, though, that the media never bothered to mention: All of these companies are publicly held. They issue stock. And when such a company incurs such massive losses they are required by law to make reports to the Securities Exchange Commission.

But none of them did.

BLACK HATS AND WHITE HATS

OK, so there are bad hackers. Mean, nasty hackers who do crash computer systems and steal things. And worst of all there are those who create viruses that have, in fact, caused massive losses to corporations and small companies, as well as individuals. Such people should be held accountable for what they do.

But they are a tiny minority. In spite of what the government wants you to believe, there are not legions of hackers plotting to take down the entire Internet, with only the heroic acts of said feds preventing this from happening. To begin with, if these people knew a little about how—and why—the Internet was established, they would know that this is absurd.

It is true that even if a handful of malicious hackers wanted to, they could cause an enormous amount of damage—far worse by several orders of magnitude than the massive denial of service attacks in October 2002 against Google and Yahoo!, in which users were unable to connect to the DNS server when core routers were shut down. And there is nothing the feds can do to prevent it.[1]

But it doesn't happen.

Maybe that should tell them something. But they are not listening.

Once upon a time, not all that many years ago, there were people who discovered flaws, weaknesses in various applications. For example, there was the kid in (I think it was) Finland who found an easy way to hack Netscape. And in so doing, the producers of the software were often grateful for pointing this out. If it were not for such people, many programs would remain unsafe (such as much of what Microsoft makes).

So draw your own conclusion about hackers, but at least here you have been able to see the other side of the issue. After all, would you automatically condemn everyone named Hannibal as someone who eats his neighbors?

ENDNOTE

1. DNS is explained elsewhere in this book, but what is interesting is that if people knew how to use Internet tools, they could do a reverse lookup and get the IP of the site they want to connect to and use it to bypass the DNS servers. You can learn about these tools in the next chapter.

Internet Tools

This chapter was originally published in the "Cyber-Street Survival" series in Nuts & Volts *magazine.*

To use some of the software in these articles, it is necessary to understand the basics of Internet addressing. Just as every telephone has a unique number, so does every computer on the Internet. This is known as its IP or Internet Protocol address and is represented as a series of digits. These numbers are what Internet equipment uses.

But humans are more comfortable with names. So, we have what are known as DNS (Domain Name) servers, which translate one to the other. If you want to use your browser to read the news, you type in http://www.yahoo.com and a DNS server somewhere, probably at your Internet Service Provider, does the translation and the connection is made. This also applies to extensions other than dot com, such as dot org, dot net, dot edu, and dot gov.

Note that IP is also used here to indicate the Internet Provider address, the particular IP that you are using when connected to your Internet Service Provider.

This chapter is about tools that can be used to gain a better understanding of the Internet, to find out who owns a Web site, or if an e-mail address is valid and more, as you will see. Some of them are built into Windows. They run under DOS, or with later versions the "Command Prompt," and can be a little confusing—DOS has never been called user-friendly. There are other individual tools that are better, available at Winfiles (http://www.winfiles.com), but the most efficient way is to use a "suite." This refers to a program that contains most or all of these tools; Sam Spade, available at http://www.samspade.org, is very good and is free.

NET.DEMON

The one I use most is net.demon from http://www.netdemon.net. It has a few things that none of the others do and it costs only $20. You can download it at their site (the filename is netdemon.exe) and install the same as any other program.

More on net.demon coming up, but first, a little more on TCP/IP. Skip it if you're not all that interested.

These addresses are usually written in dotted quad notation, as a series of four 8-bit numbers, written in decimal form and separated by periods, for example 151.196.75.10. Each number is in the range 0–255, so if you ever see something that looks like an IP address with numbers outside those ranges it's not a real address.

The leftmost number is the most significant, and the rightmost the least.

So 151.196.75.10 and 151.196.75.11 are sort of right next door to each other, but 151.196.75.10 and 152.196.75.10 are completely unrelated.

IP ADDRESS ALLOCATION

Traditionally, IP addresses were allocated to companies and ISPs in blocks.

A Class A Address Block or, less formally, an "A" Block, is a block of 16,000,000 or so (2^24) addresses from X.0.0.0 to X.255.255.255, where 0 < X < 127.

For example, the entire 9.0.0.0–9.255.255.255 range of addresses is the A Block owned by IBM.

A "B" Block is a block of 65,000 or so (2^16) addresses from X.Y.0.0 to X.Y.255.255, where 127 < X < 192 and 0 <= Y < 256.

A "C" Block is a block of 256 addresses from X.Y.Z.0 to X.Y.Z.255 where 191 < X < 224, 0 <= Y,Z <256.

(There are also D and E class addresses allocated in the 224–255 range; they are reserved for multicast and experimental applications, and you'll never see them in practice)

Traditional blocks are often described using the first address in the block, e.g., IBM owns A Block 9.0.0.0 and Cyberpromo owns C Block 205.199.2.0. Other times they may be described using just the constant prefix, e.g., net 9 for IBM or net 205.199.2 for CyberPromo.

(You'll often hear any address range from X.Y.Z.0–X.Y.Z.255 called a C Block, even though it technically isn't unless 191 < X < 224.)

BACK TO CONFIGURING NET.DEMON

At the top left corner, click File and then Options.

In the window below, you can enter an e-mail address. It doesn't have to be your real address; you

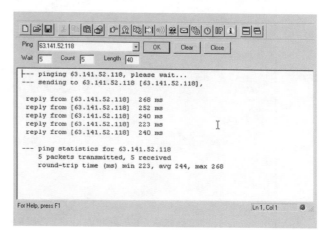

Pinging with net.demon.

can use one of the free accounts such as Hotmail or Bigfoot that you read about in Part 3. Check all of the three little boxes below. Ignore, for now, the Nameserver box. Click OK. We will come back to it.

Across the top of the screen are a number of icons: The yellow hand is Finger, then Whois, DNS, Traceroute, Ping, a funny face which is Stupid URL Tricks, an envelope that is E-mail Verify, WWW, IP, and Info. Let's have a look at these tools, but not in that order, starting with IP.

IP

Remember that the Internet uses numeric addresses (sometimes called dotted quad notation), but people are more comfortable with names. The IP tool is used to convert addresses back and forth. Click the IP button. In the window type http://www.nutsvolts.com and click OK. You'll see:

—- resolving host "www.nutsvolts.com," please wait... www.nutsvolts.com [63.141.52.118]

Net.demon has found the numeric IP address for the *Nuts & Volts* site. Now, do it in reverse, typing in 63.141.52.118 and hit OK.

—- resolving IP [63.141.52.118], please wait... www.nutsvolts.com [63.141.52.118]

Net.demon did the lookup in reverse. Try some others if you like.

PING

The icon with the yellow dot and red lines is Ping, which works something like the sonar systems used in submarines. The sub sends out the signal we have all heard in movies to see if it returns. Bounces off something. Such as an enemy sub or Charlie Tuna. Your computer does the same thing—sends out a signal to an IP address you specify and waits to see if it echoes, bounces back. If it does, there is something at that address. If it does not return, there may still be a computer there, but it is probably in stealth mode like you can see at Gibson Research.

In this example I will ping the Electronic Frontier Foundation, http://www.eff.org.

—- pinging 204.253.162.11, please wait...
—- sending to 204.253.162.11 [204.253.162.11],
reply from [204.253.162.11] 311 ms
reply from [204.253.162.11] 194 ms
reply from [204.253.162.11] 196 ms
reply from [204.253.162.11] 238 ms
reply from [204.253.162.11] 188 ms
—- ping statistics for 204.253.162.11: 5 packets transmitted, 5 received
round-trip time (ms) min 188, avg 225, max 311

Since the pings returned, something is there. Incidentally, some Ping programs will have "TTL" in the report. This stands for Time To Live. The Pings you send only last for so many thousandths of a second. As they go along their path to the destination, they lose some of that life and eventually "decay" and end.

TRACEROUTE

The shortest distance between two points is a straight line, but on the Internet it doesn't work that way. Suppose you send an e-mail to your granny, thanking her for the horrible ties she sent for your birthday. You are in Chicago; she is in Helena, Montana. The message you sent might go by way of Palo Alto, California, or somewhere in Virginia to get there. And along the way it passes through any number of locations, making what are called "hops." With Traceroute you may be able to get a list of these locations. Click the right-pointing arrow and enter a target, using either the numeric IP or the domain name.

An example: I will do a trace from the computer used to write this to a Tecra notebook computer, which is also here on my desk. But first, I need the IP of the Tecra. So, I use a tool called IP Agent (from http://www.grc.com) and get the Tecra's IP, which is 208.128.203.197. Type this into the net.demon window and hit OK and it displays:

—- looking up host 208.128.203.197
—- traceroute to 208.128.203.197
    [208.128.203.197],
30 hops max, 18 byte packets
1 [208.128.203.195] halfdome.istep.com 123 ms
2 [208.128.203.197] dial01.istep.com 283 ms
—- traceroute statistics for 208.128.203.197
2 packets transmitted, 2 received. Round-trip
    time (ms) min 123, avg 203, max 283

The numbers after the names—123 and 283—are the time, in milliseconds, it took for the trace.

What it did was send a Ping from one computer to the other through my ISP, which is http://www.istep.com, so there are only two hops: the Tecra (208.128.203.197) and my ISP (208.128.203.195). Istep is only about three miles from here.

As another example, I will use the path from this computer to http://www.eff.org, the Electronic Frontier Foundation. It is located on Bryant Street, here in San Francisco, less than three miles away, so the result should be about the same, right?

I type http://www.eff.org in the window, hit OK, and get:

—- looking up host www.eff.org
—- traceroute to www.eff.org [204.253.162.11],
30 hops max, 18 byte packets
1 [208.128.203.195] halfdome.istep.com 130 ms
2 [208.128.203.193] router.istep.com 136 ms
3 [204.153.194.149] bordercore1-istep-sf.isp.net
    135 ms
4 [205.171.37.237] 205.171.37.237 132 ms
5 [205.171.18.2] sfo-core-02.inet.qwest.net 135 ms
6 [205.171.5.113] jfk-core-02.inet.qwest.net 182 ms
7 [205.171.30.1] jfk-core-01.inet.qwest.net 179 ms

8 [205.171.5.235] wdc-core-02.inet.qwest.net 188 ms
 9 [205.171.24.1] wdc-core-01.inet.qwest.net 190 ms
10 [205.171.24.38] wdc-brdr-03.inet.qwest.net 191 ms
11 [205.171.4.70] 205.171.4.70 184 ms
12 [146.188.162.214] 129.at-0-1-0.XR2.DCA8.ALTER.NET 184 ms
13 [152.63.35.249] 152.63.35.249 182 ms
14 [152.63.3.246] 115.at-6-1-0.TR4.SCL1.ALTER.NET 243 ms
15 [152.63.48.137] 499.at-1-0-0.XR4.SCL1.ALTER.NET 183 ms
16 [152.63.48.169] 294.ATM10-0-0.GW2.SCL1.ALTER.NET 181 ms
17 [204.253.162.1] gateway.eff.org 197 ms
18 [204.253.162.11] www.eff.org 198 ms
—- traceroute statistics for www.eff.org
18 packets transmitted, 18 received
round-trip time (ms) min 130, avg 175, max 243

What happened here? 18 hops to go three miles? Look at lines 4 to 11. All of these IPs belong to Qwest, which is sort of like a huge ISP. They have their own fiber-optic cables, and many individual ISPs have their traffic passing through (in and out) of Qwest. And because they are so big they need a number of routers, each of which (like everything on the Internet) has its own IP address. So, my pings bounced around at Qwest, going through different routers until it found the right one. Then off to the next hop. If you try this yourself, you will find that some of the hops contain no information—you see only the "*"—meaning that the ping was not returned and from that you know that whatever is there is in stealth mode, if additional hops do show an IP.

VISUALROUTE

VisualRoute is another program that shows the list of hops and displays this information on a map. Using VisualRoute, the hops are from this computer in San Francisco to my dial-up connection in Emeryville, a few miles north, then back to San Francisco, then Fairfax, Virginia, Santa Clara (south of San Francisco), and finally back to EFF on Bryant Street. To go a few miles, the trace goes via Virginia, just like the e-mail you sent to your granny. In Virginia the trace goes through several routers at AlterNet, which is similar to Qwest.

You can do an online demonstration at http://visualroute.com/ by typing in the URL of the site you want to trace, such as www.eff.org, and see it happen right there. Or you can download a trial version. This is a very nice program, not only for Traceroute but also because with a single click on any of the hops in the report, you can do a Whois, which is the next net.demon tool.

VisualRoute is available at http://www.visualware.com/.

WHOIS (WHO IS?)

With Whois, you can do things such as find the owner of a Web site. This example is from the "Cyber-Street Survival" series published in *Nuts & Volts* magazine.

Click the icon with the question mark. Type in http://www.nutsvolts.com and you'll see this:

Registrant:
Nuts & Volts Magazine (NUTSVOLTS-DOM)
430 Princeland Court
Corona, CA 92879
Domain Name: NUTSVOLTS.COM
Administrative Contact, Billing Contact:
Lemieux, Larry (LL347)
    larry@NUTSVOLTS.COM
T & L Publications, Inc. 430 Princeland Court
Domain servers in listed order:
NS1.WEBHOSTING2U.NET 63.140.75.240
NS2.WEBHOSTING2U.NET 63.140.75.241
—- connection closed

The rest of the information is deleted to save space. Most of it is self-explanatory. Larry Lemieux, the editor of *Nuts & Volts*, owns the domain name www.nutsvolts.com. Greg Jacobs is at IMAGE-2020.com, which is the Internet service provider for the *N&V* Web site and also the producers of the new *N&V* site. The domain servers are the DNS you already read about; they convert from names to numbers and back so you can get connected to the site you want.

> IF YOU WERE USING A BOLT CUTTER while sizing steel rebars for a building foundation, then it would be a legitimate tool. Get caught with one while hanging out in an alley behind a jewelry store and it becomes a burglary tool.
>
> The same applications that are used for network security surveys and testing can be used to unlawfully access said networks, as you can read about in the chapter on wireless networks.

There are other things you can do with Whois. You can look up a partial address or a name to track down spammers. This is well explained at http://www.netdemon.net/tutorials/whois.txt, which was written by Matt Schneider, the programmer who produced net.demon.

STUPID URL TRICKS

Spammers sometimes try to disguise their IP address so you can't trace them. But with this tool, you can sort through the gibberish and find the real IP. One example from the Help file is http://0321.0314.0341.036/768.html. Doesn't look like the IP addresses you are familiar with, does it? Click the icon with the strange face, type it in, and hit OK. Aha! The real IP is 209.204.225.30. To see some more examples go to http://www.netdemon.net/help/urltricks.html.

FINGER

Finger is a tool used to find information about someone. It was set up in the early days of the Internet when people would write up a profile—perhaps a resume, maybe personal and contact information—that was placed at their UNIX server for anyone to see. This was called a "Plan." But it doesn't seem to be used much anymore. Some ISPs block Finger requests and advise you to contact them directly for more information. Most of the time, however, you will get "Connection Refused."

Try it with some friends' addresses. You never know what you might find. Another interesting trick is to use Finger on an IP address, not just e-mail. When you do a Traceroute, look at the next to the last listing. It will sometimes include the word Gateway. Now copy the IP of this next-to-last hop and do a Finger on it. You may see, before your startled eyes, a list of everyone who is logged on to the site. Sometimes it's a very long list.

WWW

This tool is used to connect to an IP that you think is a Web site—just enter the IP and see. If it is and it is not blocked, you will see the source code of that site—the HTML used for building Web sites. Then, you can use your browser to log on. In many cases you will see "connection refused," which could mean any of several things. You might be trying to connect to someone's personal computer, rather than a site or a server. Unless the user has Port 80 open and "listening," you'll get connection refused. Remember that because you can use Port 80 to view a Web site does not mean that Port 80 is open.

In the above example of URL tricks, we resolved to the IP 209.204.225.30. Type this in and you'll see an example of the HTML code. Now, open your browser and enter the same IP and you'll be taken to a school that teaches martial arts. Interesting, no?

INFORMATION

The icon with the yellow letter 'i' displays information about your computer.

Winsock Version: Winsock (Windows Socket) is what Windows uses to make the Internet connection. The version you are using is shown here.

Highest Version: The latest version available
Description: Description of the Winsock
System Status: On Win95.
Max Sockets: 32767
Max UDP Datagram: 65467
Socket Types: Stream, UDP, RAW
Nameserver: ns.internic.net
Local Host: mycomputername, the name you
    used when Windows was installed
IP Address: You know what that is
Resolved Host: Your Internet conection at
    the ISP

NOW, BACK TO CONFIGURING NET.DEMON

If you try to use the DNS tool, you will get an error message, Host Not Found. To make it work, you need to tell net.demon the IP addresses of the Name Server(s) that your ISP uses. Now that you have tried out Whois, you know how to find them. Click the Whois icon and enter the name of your ISP, such as http://www.myserver.com, and click OK. At the bottom of the window you will see something like this example:

```
Domain servers in listed order:
NS1.ISTEP.COM 216.200.201.12
NS2.ISTEP.COM 208.128.203.208
—- connection closed
```

Write down the IP addresses. Then click File and Options to get the Configuration screen. Check Specify and in the window to the right, enter the first IP address. Where in this example it is 216.200.201.12, you would enter the actual IP you got from Whois. Then hit Enter and type in the second IP address, click OK, and you're done.

To verify that it was done right, close net.demon and restart it. Click the DNS icon. In the window to the right of Server, you will see the first IP that you typed in: the one you got from Whois.

E-MAIL VERIFY

Some of the addresses you find in spam are forged. This tool is used to see if an e-mail address is valid. To use it, type in the suspected e-mail address and click OK. The result may be a little confusing, many lines of text with terms that may as well be Greek. What you are looking for is Recipient OK, which will be at the bottom of the screen. As an example, I will use the same e-mail address as in Finger, the publisher of 2600 magazine. Here is the result:

```
—- 10/09/00 03:17:37 Pacific Daylight Time
—- verify e-mail "emmanuel@2600.com"
—- looking up mailexchange for 2600.com
—- contacting nameserver 216.200.201.12
    (216.200.201.12)
—- found mailexchange "phalse.2600.com"
—- contacting host phalse.2600.com
    [216.66.24.2]
220 phalse.2600.com ESMTP Sendmail 8.8.8/8.8.8;
    Mon, 9 Oct 2000 06:13:26 -0400 (EDT)
> HELO istep.com
250 phalse.2600.com Hello dial04.istep.com
    [208.128.203.200], pleased to meet you
> VRFY emmanuel@2600.com
250 Emmanuel Goldstein
    <emmanuel@phalse.2600.com>
> MAIL FROM: <mdavis@istep.com>
250 <mdavis@istep.com>... Sender ok
> RCPT TO: <emmanuel@2600.com>
250 <emmanuel@2600.com>... Recipient ok
> EXPN emmanuel@2600.com
250 Emmanuel Goldstein <"|IFS=' ';exec
    /usr/pkg/bin/procmail"@phalse.2600.com>
—- verify completed
```

Again: It is not necessary to understand all of this stuff. Just ignore it. What you are looking for is Recipient OK, which tells you that this is a real e-mail address.

Now, take the e-mail address from a spam, Michelle_4025292@worldnet.att.net, and see what happens.

```
550 Invalid recipient: <Michelle_4025292@
    worldnet.att.net>
—- verify completed
```

This means it is *not* a valid e-mail address.

Secret Computer Codes:
Data Encryption

"Why would anyone want to use encryption?" the government wants to know. After all, there's no reason to unless a person has something to hide. "They must be guilty of something . . ."

First, some history, questions and answers, and then some recommendations on keeping your data secure.

People have had secrets as long as there have been people. Who knows but that some of the drawings that date back thousands of years might have been love notes from a cave person to his or her significant other or whatever they had back then before marriage ceremonies.

Perhaps the hieroglyphics inside the pyramids of Egypt were secret messages. If so, they were damn good as some of them have not been deciphered 50 centuries later.

One of the earliest, at least that could be written on paper, was the Caesar Cipher, which was a simple letter substitution: A=C, B=D, and like that. An improvement was to write on a strip of parchment wound around something such as a sword handle; it could be read only by rewrapping it on something the same size and shape.

Then there was Playfair and Exclusive-OR and many others and, of course, the One Time Pad, which you can read about in Ken Follet's *The Key to Rebecca*.

Q: I have heard that the government has computers that can instantly break any kind of code. Is this true?

A: No.

First of all, the term "break" is inaccurate, as it suggests that there is a way to decrypt any message encrypted with a particular cipher, even though the keys or passwords are different for different messages. In other words, break the cipher rather than find the key for a given message. This assumes that there are no weaknesses like a trapdoor or Trojan horse. For example, it is rumored that the encryption routine in Windows was compromised by making part of all generated keys exactly the same. It was called the Work Reduction Act.

There are various ways to go about attempting to convert an encrypted message back to readable form, called plaintext. This, incidentally, is called an "attack."

One of these attacks is called brute force, in which every possible key is tried until the right one is found, or derived. Whether or not this can be done depends on the length of the password (called keyspace) that the encryption algorithm (program) uses.

If an agency, such as the NSA, were to try and find a password, they might use something called "libraries." This would consist of gathering every single fact they could find about the person who encrypted the message(s). Not only personal information such as past addresses, Social Security number, and names of friends and neighbors past and present but also the names of the streets in the city where they went to college, their instructors' names, terms used in the subjects they studied, names of well known authorities and experts, inventors, technical terms, anything to do with their hobbies and interests (photography, flying, sailing, etc.) — literally anything they can find.

I suspect they already have special libraries stored in their massive systems deep underground in places like Ft. Meade, Maryland, and Langley, Virginia. To build a custom library for a suspect who attended Columbia University, they would punch a few keys and every known fact about Columbia would be copied to this custom library. If the suspect was known to have dogs as pets, thousands of facts about man's best friend would be added to the library.

Next, they would take all of these terms, names, and numbers and sort them into groups of the size that a password probably would be; perhaps 64-character blocks. Then, all of these blocks would be rearranged into all of their possible combinations. Spelled forward and backward, etc. If this seems like a lot of information, consider the alternative — every possible key. The entire contents of the Library of Congress might add up to a fraction of the possible password combinations. Again, the numbers here are academic.

Data runs faster in hardware than in software, so there is no doubt that the government has built massive machines, perhaps occupying enormous underground rooms in Maryland, to attempt brute force attacks, but instead of libraries, they would methodically generate every possible key combination.

This, just generating huge numbers, is a simple task and so is extremely fast. But trying these gazillions of keys isn't. Suppose it takes one of their monster computers 20 clock cycles to generate a single possible key. Well, applying that key to an encrypted message might take 200 clock cycles, as it has to make massive comparisons and compare the results to known words which, if found, would trigger an alarm of some kind. But if the original text were in another language or in five-letter code groups as used by the military in World War II, this method might fail.

Naturally, the NSA isn't commenting, so there is no way to know how far advanced they are. Some 10 years ago I wrote this:

> Suppose some agency, perhaps NSA, starts with a machine that can try one billion possible keys per second, and then parallels a thousand of these machines. All of them working together to derive the 1,024-bit private key of an RSA ciphered document. At that rate, not allowing for breakdowns, it would take about 10,782,897,524,600 years. Given that, on average, the right key would be found after trying half the possible keys, it might be derived in about five trillion years. By then the information would no longer be of any importance.

Since then, when the average home computer was a 486, technology has increased beyond my ability to put numbers on it. But the basic fact remains that it is easier to encrypt information than it is to decrypt it back to its original form, and so it is logical that by using a long enough key, it will still take the NSA or NIST (National Institute of Standards and Technology) so many years to derive the original text that it will no longer be of value.

But once again, this all depends on the use of hard-to-guess keys of sufficient length, and of course changing them on a regular basis.

WHAT ABOUT THE DATA ENCRYPTION STANDARD?

As stated above, no one has ever "broken" the DES. However, several people from, I believe it was the Electronic Frontier Foundation or Computer Professionals for Social Responsibility (CPSR), have built a special purpose device that can obtain the key used for a given message within a few days. It cost something like half a million dollars, well within the budget of any federal agency or large corporation. There is no doubt that the NSA has a similar device, but many times larger, that can find the key in a matter of minutes, if not seconds. The DES is safe enough to defeat the average person or technician, but not the government or anyone with half a million bucks to spend. Even though it is unlikely that we "average" people have anything that big business or government would take the time and expense to decode, why take the chance? The purpose of encryption is to keep your information secret. Why compromise it? The DES is obsolete.

There is a variation of DES called DESX or "triple DES" available from http://www.rsasecurity.com/. This is another very secure cipher; if interested, please go to the RSA site for details. Incidentally, it was RSA scientists who produced the first private/public key crypto-system.

WHAT IS PGP?

Pretty Good Privacy, developed by Phil Zimmermann, is a "public key" cipher, meaning that it has two keys. One you keep secret; the other, the public key, you can make available to others. If someone wants to send you a secret message, they use your public key. Only your private key can decode it. Conversely, a message encrypted with your private key can be decoded only with your public key. PGP uses a cipher called IDEA (International Data Encryption Algorithm) to encrypt the message and then the RSA algorithm to encrypt the public and private keys. To learn more about PGP, please connect to their Web site, http://www.pgpi.org/.

How secure is it?

According to their Web site, "A typical 1,024-bit PGP message would take about 300,000,000,000 MIPS [million instructions per second] year to crack, so the ordinary citizen is relatively safe off, at least for the next few decades."

Three hundred billion years for a machine running a million instructions per second. A fast Pentium IV might run about 5,000 MIPS. Some of the so-called super computers may run five million MIPS. Divide five million into 300 billion . . . you get the idea. If anyone were that interested, they would more likely find a way to steal your key, such as placing a "Key Logger" on your keyboard, or using any of the other methods of surveillance you can read about here.

If I Use PGP, What Is a Safe Password?

The last time I used PGP, the keys were generated by moving the mouse around until the random number generator produced enough digits to start with, and from there the two keys were computed. Then, the user selected a pass phrase. The best one would be a long string of random characters including letters (upper and lower case), digits, and punctuation marks. However, this is difficult to memorize and would require that it be written down, which is NOT a good idea. So you can use a phrase that you make up, something that is easy to remember but means nothing to anyone else. "I'll never forget that SumMer eveNing in PawtuckEt in 1972," or "When I wAs a cHild, I had a pet aaRdvark named Andy," or a nonsensical phrase such as "Magic mageNta martians maKe marvelous moonbeam martinis." Or something like that. But be absolutely sure that you can remember it. If you lose or forget your password, anything encrypted with it is lost. *There is nothing anyone can do to get it back—it is gone forever.*

Another thing to know before you use PGP is authentication, hashing algorithms. So please read the Help files first.

I have heard that PGP has been compromised, that Mr. Zimmermann gave in to pressure from the NSA some 10 years ago and produced the newer versions with a weakness built in, such as the Work Reduction Act mentioned above. I have no idea if there is any truth to this or not. A number of people who are supposed to be in the know suggest using the earlier version of PGP, but you will have to decide this for yourself.

There are lots of ciphers and encryption programs advertised on the Web. How do I know that I can trust them? How do I know that they do not have trapdoors or that they are strong enough?

The first thing to consider is whether the source code (the program in its "plain English" form, before it is compiled) is available. If it is not, the program should not be used. Period. This is not to say that it has a weakness or that it is not strong enough; it is to say that without the source, you have no way of knowing what it is. For all you know, this could be a Trojan horse produced in Russia. It has happened before. It could be produced by the NSA. Perhaps you are not aware of it, but the federal government has been spending a great deal of the taxpayers' money to prevent Americans from having secure encryption.

So why pay money for something you aren't sure of when you can have something that is very secure and time-tested for a few bucks or even for free?

WHAT EXACTLY IS RSA?

RSA is a public-private key cipher named from the letters of the names of the inventors, doctors Rivest, Shamir, and Adleman, and issued patent #4,405,829. How it works:

It starts with two very large prime numbers. For this example, however, I will use small numbers; "P" will be 3 and "Q" will be 11. Then, P and Q are multiplied to get 33, which we will call "N."

Next, we find a relatively prime (RP) number called "D" with this formula: D=RP to the [P-1] * [Q-1] power. Put another way, D is relatively prime to the product of P-1 times Q-1.

Now, P was 3 and Q was 11, so P-1 is 2 and Q-1 is 10. Two times 10 is 20, and a number that is relatively prime to 20 is 7. So, D becomes 7.

We now have P=3, Q=11, N=33, and D=7.

The next step is to find E, using the formula [E*D] mod ((P-1)*(Q-1)) = 1. (Mod means the remainder after numbers are divided.) We know that P-1 times Q-1 = 20 and D is 7, so E will be a number that can be multiplied by D (7) and then divided by 20 so that the remainder (mod) is 1. Such a number is 3. Three times 7 = 21. Divide 21 by 20 and you get a remainder of 1.

We now have N, which is 33; D, which is 7; and E, which is 3.

The two keys used in the public key system are made from these three numbers. One key (public) is N (33) and D (7); the other (private) is N (33) and E (3). (The "and" here does not mean the two are added; they are separate parts of the key as you will see.)

OK, now that we have a pair of keys, let's use them to encrypt a secret message. The RSA works with numbers (which in the computer program represent letters), so our secret message, for simplicity, will be the digit 4. We will encrypt 4 using the public key 33 and 3.

First step is to raise 4 to the third power, which is 64. The reason we use the third power is because 3 is part of the key.

Second step, we calculate 64 mod 33, using 33 because it is the other part of the key. So, 64 divided by 33 has a remainder of 31. The secret message 4 has been encrypted and is now 31. We send this encrypted message to another person. Then, he will use his key and try to decrypt the 31 back to its original form, which is 4.

First step is to raise 31 to the seventh power, because 7 is part of the other key. Take 31^7 and get 27,512,614,111.

Next step is to divide 27,512,614,111 mod 33, which is 4. The secret message has been decrypted back to 4.

Another example using 5 as the secret message and the same key, 3 and 33. Raise 5 to the third power and get 125. 125 mod 33 is 26. The encrypted message is 26.

Now to decode it using the other key, 7 and 33. The 26 is raised to the seventh power, which is 8,031,810,176. That number mod 33 is 5, which we started with.

If you try this using other numbers for the keys, you will see that it doesn't work: What is encrypted with one key can be decrypted only with the other key.

That's how the RSA cipher works. Now, imagine that we started with two prime numbers that were 300 digits in length. To decrypt a message with such long keys involves some very heavy number crunching, called something like "factoring the product of long prime numbers."

The last I heard, several years ago, the longest number that had been successfully factored was based on an 80-digit prime. A 300-digit prime is something for the generation of computers in the distant future.

WHAT IS STEGANOGRAPHY?

The word comes from the Greek *steganos*, meaning to hide or conceal, and *graph*, or picture. It is, among other things, the technique of hiding something within something else—of concealing a secret message inside a picture. Read more about it and get on a mailing list here: http://www.petitcolas.net/fabien/steganography/

If you use either DESX or PGP—and use it right—you have little to be concerned about as far as anyone deriving the key(s) through any method known to big business or government. Unless you are a wanted criminal or suspected terrorist, who is even gonna try?

Yes, there is the rumor. That Phil Zimmermann who wrote PGP (using RSA Data Security's algorithm described above) weakened it similar to the alleged weakening of Windows. I don't believe it, since the source code is published. But if you have any doubts, use an earlier version.

Mr. Zimmermann's site is here: http://www.philzimmermann.com/EN/background/index.html.

Part Five

It's a Wireless World

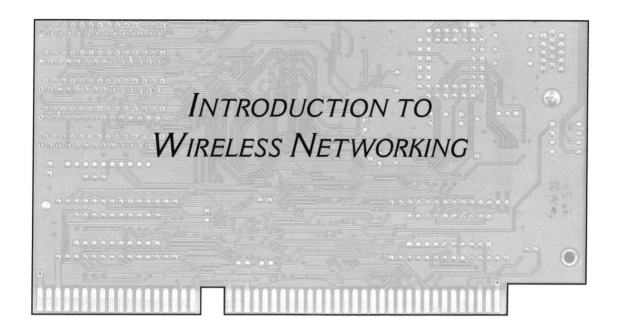

Introduction to Wireless Networking

This opening chapter about wireless computer networks is written for beginning and intermediate level users, starting with the very basics and moving into more detail so that when you have finished reading it, and have done the exercises (there might be a pop quiz now and then), you will know a fair amount about 802.11, or wireless fidelity (known as Wi-Fi).

You will know how to get started with your first Wi-Fi card for a portable computer and how to get logged in to the access points (AP), or "hot spots," at local Internet cafés. From there, if you want to move on, you'll learn how to increase the range of your card with the right antenna and how to monitor various APs and see what is happening there. Under some circumstances, you will be able to monitor an AP and a particular user who is connected to that AP and see what he is seeing on his screen, wherever he is. You can learn about the many programs, called "packet sniffers," that can be used to monitor APs.

Should you decide to set up your own AP, to be used for your own little home network and be able to have several computers using the same DSL connection, you will find the answers here. And even more important, you can learn how to make your network secure, so that the people across the street or war drivers are unlikely to get into your computers or use your AP to get a free Internet connection. Unless you choose to let them.

Now, there are hundreds of Web sites that have information about all these things. You can search and, over a period of time, learn much of which is in this chapter, but it isn't easy figuring out what goes with what and how this or that works. The many newsgroups, for example, assume that you already know the basics, so they aren't included. It is considered lame to ask fundamental questions, and if you do, you may be ignored or flamed. That's how it works.

Likewise, there are dozens of books with which you can learn all that you need, but none of them that I have seen boils down a massive amount of information and then sorts it out into an easy-to-follow tutorial. The purpose here is to do just that, and in some cases present step-by-step instructions. *If you are totally new to wireless*

This old Pentium II runs both Windows 2000 and Linux Red Hat 8.0. This will be referred to in the text as the "left" computer, or "box."

The main computer (shown here on the right), a desktop with the 17-inch Sony monitor, is the "center" or "middle" computer. The small black box on top of the speaker is a Siemens DSL router, AP, and firewall. On the left is the Pentium II running Linux. The scope is a 15 to 60 zoom made by Parks.

Third down is a Compaq Presario notebook computer on the right, which I will call "right" or just Compaq. I do not recommend Compaq for many reasons.

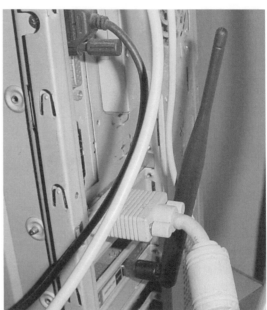

A LinkSys WMP11 PCI wireless card with a stock antenna on the Linux box, the same antenna as seen on the router.

A 14-dB Yagi 2.4 GHz antenna.

The 24-dB magnesium "grid" antenna. Mounting it as shown proved to be easier said than done until I found the tripod at Goodwill for 10 bucks. It has a flat aluminum plate on top and I took it, along with the "L" bracket that holds the antenna, to a local welding shop where they heli-arced it in place. For such a directional antenna, precise aiming is necessary.

computing, you have no idea how much time this material will save you. Not to mention a considerable amount of money. There are many brands of computers and wireless cards and APs and cables and connectors as well as tons of programs. And even though 802.11b is an industry standard, not everything works together.[1] Some programs work with some wireless cards but not others. Some cards have external antenna connectors, but they are not all the same. So, an antenna cable might fit and it might not.

Yes, it can be confusing, so read on, and while it may seem like I hop from one subject to another, hang in there, and I will put it all together. So, again, when you complete these chapters, you will have a good, solid working knowledge of wireless computer networking.

In researching this, I used the equipment pictured in this section, and at various places in the text I will make reference to it.

OK, let's look at some of the mistakes, often expensive ones, a beginner might make and how to avoid them.

ENDNOTE

1. There's more information about the differences between 802.11a, b, and g in coming chapters and in the Glossary.

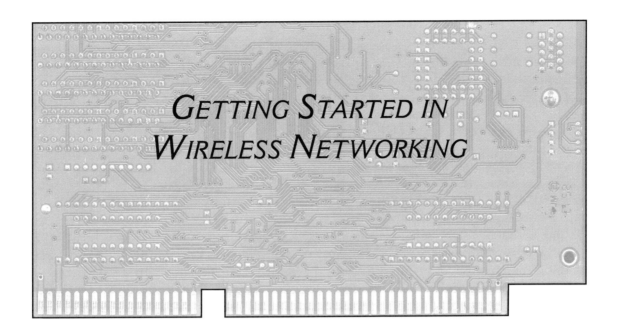

Getting Started in Wireless Networking

ONE DAY YOU WAKE UP AND DECIDE THAT YOU ARE INTERESTED IN WIRELESS computing. Or maybe you have thought about it for a while but suddenly made the decision. You already have a portable computer so you run down to Circuit City and buy a wireless card.

Once home you install the drivers from the included CD, pop the card in, and with a little luck everything works.

On your way back from your latest excursion of throwing money into the black hole that is computing, you happened to see a wireless Internet café. So you dash back there, ready to experience wireless for the first time. You go up to the counter and get some coffee and ask the person how to get connected. This may be automatic at free access places, or you may need to buy a little card explaining that all you have to do is type their address (URL) into your browser, such as Opera, and just like that, you are connected. Probably.

So, you get on Yahoo! and read the sports scores and whatever else is of interest, thinking, "Wow, this is great! Look, Ma: no wires." Yep, wireless is wonderful. So after six cups of house blend, which isn't known for being weak, you are wired as well as wireless. Time to head for home and a beer.

DO TRY THIS AT HOME

Settled at your desk, you think, "Wow, wouldn't it be great if I could do this at home?" You could take your portable (sometimes called a "laptop") out in the backyard and get some sun while telecommuting (or playing hooky from work), or even take it over to the Smiths' across the street to show off. (Not knowing that they are way ahead of you and already have their own network.) But, alas, nothing is within range. Now, being a reasonably astute person you realize that what you need is a better antenna. Much better, maybe.

On the way back to Circuit City to blow some more of your hard-earned cash, you are watching for Internet cafés near your home. And oh, lucky you, there is yet another one down the street. Betty's Bytes and Bagels. It's been there all the time, but you just never noticed it.

So you grab a spiffy new Belchfire "extended range" wireless antenna off the shelf, fidget while the clerk runs your credit card, then dash home, rip the box apart, and get ready to plug it in. But . . . there is no place to plug it in. The card you bought doesn't have an external antenna connector. Matter of fact, very few do and almost no retail stores sell them—not Circuit City, CompUSA, Office Depot, or any of the few remaining independent stores. You will be very lucky to find one. (Read on.) So you take the antenna back to Circuit City where they credit your card, no problem. CC is really great about that, but meanwhile, what the hell do you do now?

Fortunately, there is a Surf and Sip in your area and you have discovered that they sell the Senao card with an external antenna jack. So, you grab a cab and hustle over and get one. Oh joy! You get home all ready to try it, and that's when you remember that you returned the Belchfire antenna. So you hustle back to get it and now, you say to yourself, you are ready.

Nope. You didn't look at the little jack on the card and mentally compare it with the plug on the Belchfire. Doesn't fit. The antenna uses a reverse SMA connector, and the card uses an MMX. Now, you are really getting irritated. What you need is an adapter called a "pigtail." Circuit City doesn't have them and, to repeat myself, neither do any computer stores in San Francisco. Perhaps in your area this is different. Fry's has them, should there be one nearby. You can call different stores but likely will strike out. As a last resort, before you decide to have one shipped, you could spend $40 for an SMC Networks Home Antenna. The part number is SMCHMANT-6 and it has the exact pigtail you need.

Yeah, now you have two antennas. Well, that ain't the end of it; you'll likely buy more in the future.

Installing the Senao requires only that you place the CD in the drive, wait for the menu, and then follow the directions. Now, you're all set. Point the antenna in the direction of that new café and see if you can get connected. The SMC is good, I like the one I have, but with 6-dB gain, unless you are in direct line of sight, you probably won't get connected. If you can elevate the computer, such as by moving it to an upstairs bedroom or maybe sitting on the roof, this will improve your chance of establishing a connection.

Now, back before you were spending the rent money on cab fare, your idea was to be able to use your notebook computer out in the backyard or anywhere else around your house. Or the Smiths'.

You ask the salesperson, with whom by now you are on a first-name basis, and he advises that you need something called an access point—an AP. And they just happen to have one on sale, a Siemens. Ah, you are back to being excited again so you buy it—a combination AP, router, and firewall. It even has a network printer parallel port so you can use the printer from any computer that is part of what will one day be your very own home network.

> MOST OF THE CARDS, PCI, that plug into a slot on desktop computers—or PCMCIA for portables—are for 802.11b, which transmits on 2.4 GHz. (PCMCIA, or Personal Computer Memory Card International Association, is an organization that has developed a standard for PC cards.) There are some for 802.11a, which seems to not be used much and works on 5 GHz, and then there is 802.11g, which also uses 2.4 GHz but transfers data much faster. Then there are cards that may work on b and g, or all three. Before you decide which one to buy, please read the rest of the wireless chapters.

So again you reduce the packing material to confetti getting it open and look for the manual. There isn't one. There is a little instruction sheet explaining that the manual and setup files are on the CD, which you pop into the computer. And then it gets complicated. The manual for the Siemens is good, but it is not excellent; for a total beginner it is a little tricky, but if you hang in there, in a couple hours everything should be up and running. And then you make an interesting discovery. The little vertical antenna that comes with the Siemens has the same connector as the Belchfire, so you can use it with the AP and you can use the SMC with your laptop.

OK, enough already. It doesn't *have* to be that difficult, but it can be. The point is that before you decide to give wireless computing a try, you really need to determine two things. How far do you want to go with this, and to that end, what exactly do you need—what works with what? Computer. Wireless card. AP. Antenna. Cable. Pigtail. So, in this chapter, let's have a look at making the decision and take it from there.

Now, if, like at the beginning, all you want to do is go to a Wi-Fi café, down the street—or wherever—with a notebook computer and be able to get Internet access, then all you need is the card. In most cities large and small, there are APs you can connect to, including private systems at some hotels and airports as well as many independents. So most any card will work. But there will likely be times—especially if you travel a great deal and sometimes off the proverbial beaten path, when these APs may not be as close as you wish. You might be in a hotel room where there is no AP, and though there is one halfway down the block, your card just can't make a reliable connection. So, you just take one of your antennas with you, or you can get another card that has the antenna built in. Senao has one, and there are many others available.

HOME NETWORK

You now have a home network, and with the Siemens router/AP you can connect as many computers to it as you want, all of which will have Internet access. This, incidentally, is called "infrastructure." So, just as with the free wireless café down the street, other people can connect to your network and use it for Internet access the same as you. Except that you are paying for your DSL connection and they are not. Later on we will read about how this works and what you can do about it.

Infrastructure is one mode. The other is peer-to-peer. A peer network is where two computers connect directly to each other without going through an AP. The software that comes with your card(s) will explain how to set them in either the AP or peer mode. With both computers using the same channel and a few settings made in your card's software, you will be connected. Now, subject to spending more money, we are ready to begin our fascinating exploration of the Wireless World.

Next, a few words about the laws that may apply to wireless network monitoring, and then on to some reviews of software, with which you can do many things.

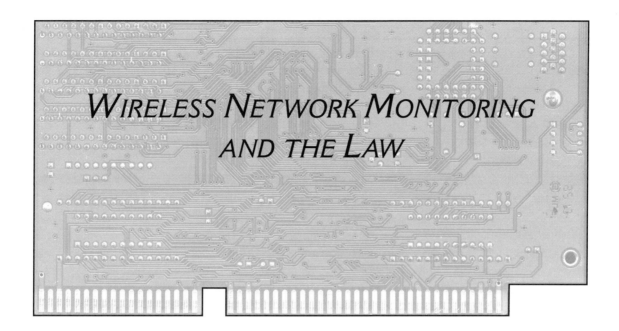

Wireless Network Monitoring and the Law

Radio waves, signals of all types, cover the entire surface of this planet. In a metropolitan area, there are so many of them—so many thousands of them—that were they to become visible, they would blot out the rays of the sun. Even if you live on a desert island or in an ice cave somewhere in Antarctica, you are inside a sea, a web, of radio signals; they are that pervasive. They cross your property and go through your house and through your body.

There are people who say that since these radio waves pass through my home and my body, then I should have the right to know what they are and what they are sending—what information is being transmitted. It doesn't work like that.

To stretch the analogy a bit, consider that any time an electrical current flows through a conductor, a wire, it radiates a weak signal. Such as the telephone. This is how inductive phone taps work without needing a physical, electrical, connection to the line. And if you live in an apartment building, then a tiny bit of that electrical energy from your neighbor's telephone may enter your living space. But obviously you do not have the "right" to monitor his phone calls. (Even though there are times you wish you could! C'mon, admit it!)

The same thing is true, to some extent, with wireless networking.

You have a portable computer with an 802.11b PC card. You go to an Internet café, and you are instantly tuned to their AP on whatever channel it happens to operate. Probably.

But suppose there is some big corporation in the same building, up a few floors, that has a wireless network. Because of the way that radio waves behave—especially at microwave frequencies, which is where 802.11 systems operate—instead of seeing on your screen a list of the kinds of coffee you can order, you see pictures of naked people of whichever sex, having whatever kind of sex because the middle managers upstairs are goofing off and browsing porno sites on company time. Or maybe you see a confidential message sent from the CEO to the chairman about a merger that they would rather no one else knew about.

> SOMETHING ELSE TO THINK ABOUT. What you connect to at the Internet café may not be what you thought it was. A wireless hacker worth his salt might have set up a "copy" of the site you intended to connect to. For example, T-Mobile. Oh, it will look just like the real thing, but it is not. You have been "wire-jacked," to coin a term. Meanwhile, the hacker is logging your user name and password and now has access to your account. You can get more details on this subject on the *Internet Edition of Electronic Surveillance and Wireless Network Hacking* Web site at http://www.fusionsites.com/dbm2.

Hey, all you wanted to do was go have a cup of house coffee, check your e-mail and the stock market.

You might also "see" an AP that is using WEP encryption, and while I don't know the law as to what I have written so far, I do know that it is unlawful to intercept any radio transmission that is encrypted or scrambled.

So what do you do? There really isn't anything you can do; your portable computer has connected to the strongest signal that it could detect, and that's what happens unless you make the effort to connect only to what you reasonably believe you are allowed to. And how many of the millions of Wi-Fi users do so? Or even know how?

Using, monitoring, sniffing, and hacking are Wi-Fi issues in which I believe the laws are vague and difficult to understand. So I suggest that you consult an attorney before you use the software reviewed here, assuming that you can find a lawyer who knows about wireless networking. And you can check out the Electronic Privacy Information Center (EPIC) and the Electronic Frontier Foundation (EFF) and other related sites.

But last and most important, consider what you could do with the information you are able to intercept, such as learning inside stock market details, pre-merger announcements, personal details of people's lives, and whatever else. Try to use it to your advantage and you may be in for the hassle of your life.

Not only are there old laws such as the Communications Act of 1934 and the Electronic Communications Privacy Act, but also remember we live in a new era, post 9-11, where we have Total Information Awareness and Homeland Security and other laws that We The People do not even know about. Keep this in mind while learning about wireless networking.

And setting up your own network.

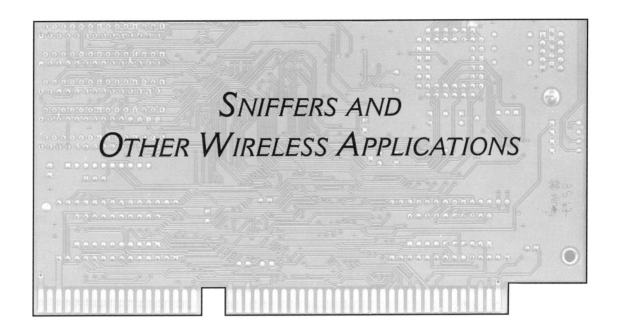

Sniffers and Other Wireless Applications

Sniffer: A program and/or device that monitors data traveling over a network. Sniffers can be used both for legitimate network management functions and for stealing information off a network. Unauthorized sniffers can be extremely dangerous to a network's security because they are virtually impossible to detect and can be inserted almost anywhere. This makes them a favorite weapon in the hacker's arsenal. On TCP/IP networks, where they sniff packets, they're often called packet sniffers.
— http://www.webopedia.com

OK, HAVING REVIEWED THE BASICS, LET'S LOOK AT SOME OF THE SOFTWARE that is available.

There are dozens of applications that can intercept wireless computer signals. Some are free, some are very expensive. One that I have read about costs more than $10,000; it is not reviewed here for obvious reasons. And in between are quite a few that the average person or small business can afford.

Some programs or apps (applications) will just make a list of the APs within range. Some can be used to intercept the data, the "packets" that are passing through a given AP, assuming it is in range: Being able to detect, or "see," the signal and connecting (associating) are not the same.

Some APs will decode, display, and store the data they see. With some you will see only raw data that means nothing. Some will convert the data so that, to a certain extent, you can see what someone else is seeing. You monitor an AP where someone is using it to connect to, say Google, and make a search, and you will be able to follow them along, again, seeing on your screen what he is seeing on his screen. In real time.

Fascinating.

Some apps work in what is called radio monitor mode (RMM). This means that the software controls your wireless card and turns off the transmit function. No one will know that you are monitoring him. One such app, CommView, is reviewed extensively here; it and how it works is very much the basis of this chapter. Most APs do not work in RMM mode, so it is possible that while you are monitoring someone, he will know this is happening. (More on this later.)

I have personally used and will review or comment on the following apps:
WinPcap
NetStumbler
CommView Wireless
Ethereal
Auditor
Knoppix
Kismet
Wireless cards

WINPCAP

WinPcap is Windows Packet Capture. Some of the programs reviewed here require it and, depending on the app, you might not be aware of this and so not understand why the program doesn't work. So it is a good idea to first get and install it. You can get this free program here: http://winpcap.polito.it/news.htm

NETSTUMBLER

NetStumbler is an industry standard and is absolutely indispensable for anyone who wants to do more than read the sports page at Betty's. The first step is to go to http://www.netstumbler.com/ and download NetStumbler. It runs on Win 98 and 2000 and other versions, and installation is painless. Make an icon to get it started if you like, reboot your computer, and start your card drivers (in case they don't start automatically) and then NetStumbler.

When it starts, you will see a screen similar to the one pictured in this section, except that you probably will not see as many listings. As you can see in the screenshot, it lists every AP as well as wireless cards in ad-hoc mode that can be detected with your PC card and antenna combination.

On the left is a list of signals which, when clicked on (see image of graphs), will show a graph of signal strength. The colored circles also indicate signal strength. Green is strong and means you will probably be able to connect. Red is very weak, and yellow is somewhere in between. But these colors are not absolute; you might connect on red but not on green. As the program scans, the dots will change to gray except for a very strong or weak signal, in which case the color does not change that often. Only a very strong signal will have a green dot that does not change.

Next is hexadecimal code (a system of counting based on 16 instead of 10). This is the "MAC," meaning Media Access Control, which is a sort of serial number burned permanently into network cards, wireless or wired, as well as APs and other networking devices. After that is SSID, Station Set Identification, which is an arbitrary, name you can give an AP. It is also optional; an SSID is not required for the AP to function.

Following that is the Name category, which more than likely will be blank except for Internet café type places. It, too, is arbitrary so you can name it anything you want—"My AP," "My Network," "Joe and Linda's," whatever.

The "Ch." refers, of course, to the channel that the AP is using. In the United States we have 11 "B" channels and others for "A" and "G." Other countries vary.

The Vendor is the manufacturer of the AP, and lastly, for now, is the notation: AP or Peer-to-Peer. If you see this, it means you are detecting someone's actual card, which can mean that whoever owns it is using their card—and probably a directional antenna—to look for APs in their area. Your area. Otherwise it is probably an AP but could be a wireless router or switch.

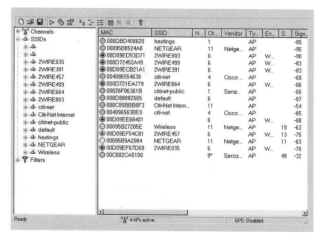

This is what you will see when NetStumbler starts.

There may be a large number of APs appearing on your screen. There may be none. So you can move the antenna around to see if you can find one or more.

Now, once you have found an AP, look at the pane on the left. Channels, SSIDs, Filters.

Click the box to the left of SSIDs, and a list of them will open. Click the "–" box and it opens, showing the SSID. If there is a little padlock in the circle to the left, it means the AP is using WEP; it is encrypted. Find an SSID without the padlock and click the MAC, and the main screen changes to a graph showing how strong the signal is. The higher the colored bars—red, purple, and green—are, the stronger it is.

Now, it is beyond this chapter to get into a detailed discussion of signal strength. What matters is what you are able to detect. Later, we will get into making a connection and being able to use one or more of the APs you see in NetStumbler and, through them, get Internet access. And, of course, by understanding all this, you will be better able to learn how to make your own AP, if you decide to set one up, secure against others who try to use it for their Internet access.

On Being Detected

At this point you may wonder, if I am using NetStumbler, can the APs that I see, see me? Do they know—can they know—that I am monitoring them?

The answer is not a simple yes or no. Technically, yes, it is possible. NetStumbler sends out a signal with some text within it that can be detected if someone is looking for it. Normally, if an AP is for someone's personal use or perhaps for a small business where they don't have an expert security consultant available, then this is very unlikely.

If said owner is also using NetStumbler, then they may see you and know that "someone" is listening, even though he probably has no more idea who you are than you do about him. In this case, he will be seeing you in which mode? This is your first mini-quiz. Hint: He will be detecting your wireless card.

Suggested Reading

There is an FAQ at the NetStumbler site that explains in detail all that you see, which you should read. The SSIDs that have a green dot, indicating a strong signal; the Sign column on the right, which is a measurement of signal strength; and the "S" column, which is the SNR or signal to noise ratio, are all important to understand. You may see a very strong signal, but if there is a great deal of noise present, you may not have very good reception— you might not capture intact packets of data.

The lower the Sign the better, and the higher the "S" the better.

Generally speaking, if the Sign is in the 60s or 70s, you should be able to monitor, capture data from that AP. The 80s are iffy, and anything in the 90s will probably not be captured at all. It depends on the card and the software that drives it. CommView, which is reviewed here, will detect weak signals, but if below a certain level will not capture packets. The Senao card is a little more sensitive than the Orinoco and, once again, the antenna is important.

Something else, for future reference: The numbers you see—Sign and "S"—are not necessarily the same as you will see in other sniffer applications, such as CommView. We'll see that when we get into the review.

NetStumbler, again, is one of the most important applications you can have for exploring and learning about wireless networking, but unfortunately, it does not work with all wireless PC

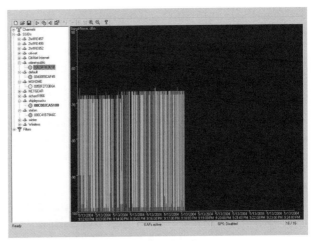

Signal strength displayed with NetStumbler. The higher and taller the green lines are, the stronger the signal. The black spaces between the green indicate that the signal has been temporarily lost.

cards. Most, but not all. I have tested it with Senao, Proxim, Orinoco Gold, and LinkSys WPC55AG. It does not work with the LinkSys PCI WMP11.

COMMVIEW WIRELESS

If you happened to read the original series of "Cyber-Street Survival" articles, you may remember that I had much to say about the wired network version of CommView. An excellent program. The same is true of the wireless version. Of all the programs I have reviewed, and there were many, I consider CommView to be the best, all things considered. So I will be using CommView as an example in the chapters on wireless networking as well as a comparison to other programs.

What's so great about it? Features, ease of use, excellent Help files, powerful Rules or "filters," and tech support, yes, but especially the ability to operate totally silent, in Radio Monitor Mode, as mentioned earlier. This means your wireless card does not transmit. So, not only will no one know you are using it—there is no signal through which it can be detected—but also if you are using a directional antenna, you will not cause interference to any other wireless network.

To repeat myself, this is *important*, as among the many wireless users out there are hospitals that depend upon the network for important patient information. And, as I have read, this can include monitoring systems such as in intensive care units.

If you decide to try CommView, you can download the trial version from http://www.tamos.com. There are some limitations to the trial version, but it is not "crippled"; you will be able to use the demo in its many features, but only every other packet of data will be displayed.

First, search their site to see if CommView is compatible with the Wi-Fi card you are using. Then start the download.

Installation

CommView installs the same as any other program; just run the .exe file and follow the directions. It is suggested, but not required, that you install it in the default directory which is C:\program files\commviewwifi. Once done, start it up and you will see a screen with instructions on configuring your card. Scroll down to the bottom of the screen and check the appropriate box to tell it whether you already have the drivers for your card installed, or you do not. If this seems a little confusing, and it might, go back and start over and read carefully what is written and you should be able to get through it OK.

If you continue to have problems, open Control Panel from Start/Settings, System, then Hardware and Device Manager. Look for the little green icon that says Network Devices. If there is a yellow question mark over it, there is an installation problem.

Right-click the icon and click Properties, then Driver and see what is there. There should be a note that the driver is TamoSoft. If not, delete it and start over.

Note that if you already have drivers installed for a different card, when you install CommView and later want to switch back to the first card, you will need to go back to Control Panel and click "Install one of the other drivers." Then select the one for the other card.

When you start it up for the first time after driver installation, click on the triangle button in the top left corner to start scanning and you will see a screen like this one:

This is the first screen you'll see upon starting CommView after driver installation.

Select which bands you want to search ("A," "B," or "G," depending on the card you have) and which channels. Turn off (don't check) Reset data after each cycle, and don't check Hide wired hosts. If this is checked, you may miss capturing some APs. Now, click on Start Scanning and let it run for a while, observing the SSIDs that appear in the left pane.

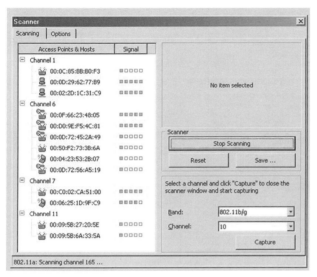

You can watch the SSIDs appear as CommView runs.

What Is All This?

If you are in a place where there is a lot of wireless activity and if you let CommView run long enough, you may well be surprised at the number of SSIDs that appear on the screen. As an example, I let it run overnight and on channel 1 are only five that are APs (icon with the "rabbit ears" antenna), and 86 others that are either "Not Wireless Host" (icon without rabbit ears), which are computers that are part of the network, but are connected with CAT-5 cable. The third icon, a portable computer, is a Wireless Host connected with a Wi-Fi card to one of the APs.

This can be confusing as you may first wonder what kind of AP can have so many wired computers. A fairly big corporation might have so many, but in this case it is Citi-Net, a company that provides wireless Internet to apartment dwellers—and this neighborhood has dozens of apartment buildings.

But that's only part of the answer. If you scroll through the list of all channels, you may find that the same MAC is on more than one channel. This may be caused by multipath distortion as described in the chapter on wave propagation. It is also possible that some of the MACs are seen on more than a single channel because the edges of the individual bands overlap slightly. Finally, there are sometimes ghost images that somehow just appear out of nowhere. This is complex and there is no need to go into it here—you can always read some of the books listed in Appendix B, particularly from the ARRL, if you want a better understanding of wave propagation. For now, let's go on with CommView.

Once there are some APs listed, select the channel you want to monitor, at first perhaps trying the channel that has the most SSIDs. Click Capture to start capturing packets; you will see a screen like this one:

Here's what you will see after clicking on Capture.

Note: For this example, I have captured packets to and from the Left computer that has the LinkSys PCI wireless card and the Siemens router/access point. This is within my own network, and I gave myself permission to do so. Observe that under MAC addresses are LinkSysPCI (MAC 00:06:25:1D:9F:C9) and MyRouterAP (00:C0:02:CA:51:00). These are aliases. With CommView you can assign an alias of your choosing for MACs

119

and IP addresses. This makes it easy to keep track of what you are seeing and has the added benefit of letting you quickly spot any new ones.

Click on any of the listings in the top window and you'll see the text that is in that packet. What you see above is an e-mail capture using Pegasus, a free e-mail program and one of the best. At the bottom is the sentence starting with "This is an example..."

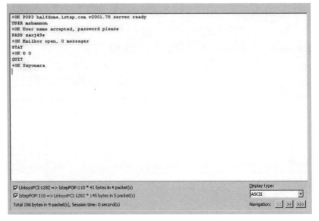

This screen has captured the password for someone's e-mail account.

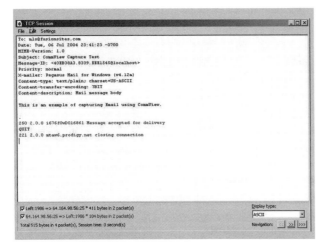

This is an e-mail capture using the free e-mail program, Pegasus.

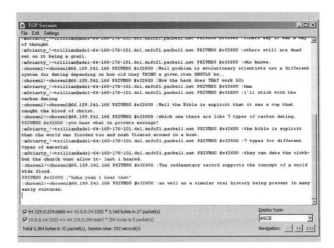

Here is another example of capturing wireless packets. This is the IRC channel #SF2600 on EFnet. We were yakking about the book The DaVinci Code.

Right click on the line and then Reconstruct TCP Session and a new window opens: This shows the message in plain text without the control characters and formats it for easier reading. If I back up a few lines, the screen would show my login name and password. In this case, the port being used was 110, which is POP or Post Office Protocol.

If it had been Port 80, which is HTTP as in browsing a Web site, there might have been graphic images included in the packets. At the bottom right is Display type:. Scroll down to HTML and click it, and as above, a new window pops open in which will be displayed the graphics. With the right pointing arrows, >> and >>>, you can follow the packets, and to some extent you can follow along and see what the person who made this WWW web site connection was seeing.

OK, at this point you have a good basic understanding of how CommView works, and with that you will have a good start on how other wireless packet sniffers work. But that's just the beginning.

The CommView Capture screen is in three parts, and you can use the little dots and bars at the extreme bottom left corner to arrange how they appear. You can have all three parts stacked vertically as in this screenshot, or the bottom part can be on the left or right of the screen.

It seems easier to start with it on the left. That way you can slide the divider bar to the left so that you see only the top and bottom frames. We can get into what information is on the left frame later, but it is rather technical and you may not even want to use it; it depends on how far you want to go with this versatile program. Now, please click Help and read Using the Program. Read it carefully, and while you probably will not absorb all that is there, it will help you understand what you see on the screen.

OK, now look at the lines of text and scroll through them, observing the different IP addresses and MACs. Again, depending on how much wireless traffic there is where you are, there might be thousands of lines from hundreds of different APs. With hundreds of MACs and IP addresses. Any number of people connected to the various APs, reading the comics, checking the stock market, sending "confidential" e-mail . . . CommView is logging everything that is being transmitted within the channel you selected. It can be mind-boggling as you look at the very bottom of the screen and see that CommView has captured several million packets.

Filters

In order to zero in on only the specific information you want, you have to learn about filters, or "rules," that control what data gets through and what is blocked. (Actually, CommView captures all of it; the filters control what appears on the screen and lets you save only the packets you want.)

Filters can be the most confusing aspect of wireless networking. Some applications, sniffers, have simple "off-or-on" filters and don't clearly show what is passed or what is blocked. Some don't actually have filters built in, so you have to create (write) your own or copy those someone else has written. And for a beginner, that would require a fair amount of searching the Web. And then, the syntax of filters for one program isn't necessarily the same for another. Capsa has a nice filter set, and Ethereal also, although it is not as easy to understand.

CommView has the most versatile set of filters of any program I have ever used. So let's get started.

At the top of the screen are three buttons, "D," "M," and "C." They stand for Data, Management, and Control. And if you click the Rules at the very top, you will see Ignore Beacons.

For now, engage only "D" for Data and don't check Ignore Beacons. This means that, so far, you want only packets of data that may contain useful information to be displayed in the bottom part of the screen. Once you have done this, clear the packet log to get a blank screen, and watch what is happening. Pause and examine some of the data by selecting one line in the top window and viewing the contents in the bottom window. You might find something interesting, but for now, let's continue with using filters.

- Beacons—All access points ship with a wireless beacon signal so that wireless PCs can find them. In effect, the beacon signal is shouting every tenth of a second or so, "I'm here! Log on!" With Ignore Beacons not checked, you will receive, depending on how many APs are operating on the channel you are monitoring, many thousands of packets that you do not need. At least not yet.

Now, click on Rules and look at the tabs on the left side of the screen.

- IP—Internet Presence of a particular computer, or the IP that a computer is connected to. Here, you can pass or block IPs.
- MAC—As explained elsewhere, MAC is Media Access Control, a hexadecimal number that is unique to every computer network card (NIC) as well as other networking equipment.
- Ports—A port is like a door or portal through which information is sent and received in a computer. There are 65,535 ports available but only the first 1,000 or so are actually normally used, that is, up to 1,023. Beyond that they are Registered (1024–49151) and Private (49152–65535). There is no need to study this long list of ports as only a handful apply to filtering in wireless networking. Click on those that you want to pass or be blocked. Port 80 is Hypertext; the World Wide Web. Port 25 is Simple Mail Transfer Protocol, used for sending e-mail through a server. Port 110 is Post Office Protocol, for receiving e-mail. Others are FTP, Telnet, IRC (Internet Relay Chat), and on and on.
- Protocol and Direction—This is advanced and will require some study that is beyond this chapter, with one exception: Click on ARP to avoid having the screen flooded with useless characters.
- Text—Text is just what it implies. Enter a string

of characters and select whether they are to be Captured (displayed on the screen) or Ignored.
- Advanced—Here, you can make a list of APs by their MAC (based on their SSID) and individually select which ones are to be Captured and which are to be Ignored. In this neighborhood there are, as previously mentioned, a couple hundred APs.

Many of these APs are from something called Citi-Net, which is a company that provides wireless Internet to the tenants living in the buildings that they own. They operate on several channels, so there is a great deal of data passing through this sophisticated system. Most of it control and management. So, to eliminate traffic from this organization, I wrote these advanced filters.

Block Citi=not((smac=00:0C:85:BB:B0:F3 or dmac=00:0C:85:BB:B0:F3))
Block Citi-Net Wireless=not((smac=00:0C:CE:0C:E9:2A or dmac=00:0C:CE:0C:E9:2A))
Block CitiNet=not((smac=00:02:6F:04:77:5B or dmac=00:02:6F:04:77:5B))
Block citi-net-public=not((smac=00:02:6F:05:EF:72 or dmac=00:02:6F:05:EF:72))
Block citi-net=not((smac=00:0A:41:7D:43:A6 or dmac=00:0A:41:7D:43:A6))

The first filter, "Block Citi," is just one of the SSIDs they use. "Not" obviously means do not display these packets on the screen. Then, smac is source and dmac is destination

So what it all comes down to is that with the use of filters, you can let every packet that is being broadcast from all of the IPs that are within range of your computer be displayed on the screen and/or saved as a file, or you can narrow this down—fine-tune it. Suppose, for example, you are an upper-level manager at a company that has hundreds of employees and dozens of APs. You have reason to believe that some workers are browsing eBay on company time, and you want to find out who they are.

You can start by running CommView to capture APs on a particular channel and then use the Text filter to set off an alarm whenever "eBay" is captured, and within the captured data will be the MAC of the computer that it is being sent to. And that employee is then invited to an interview with the office manager.

Advanced filters can get much more complex than just blocking certain MACs from Citi-Net. For example:

((sip from 192.168.0.3 to 192.168.0.7) and (dip = 192.168.1.0/28)) and (flag=PA) and (size in 200..600) //

This one captures TCP packets the size of which is between 200 and 600 bytes coming from the IP addresses in the 192.168.0.3 – 192.168.0.7 range, where destination IP address is in the 192.168.1.0/ 255.255.255.240 segment, and where the TCP flag is PSH ACK.

Again, it is complex and requires some study, but the possibilities are virtually limitless in what you can do to narrow down incoming (Source) and outgoing (Destination) packets of data. And you can save any rule set you have built—as many as you want. So, if you have a set for watching one particular employee, you can save that set as the person's name, or save a set for a specific MAC, and then load, on the fly, the set you want to use at any given time.

WEP and WPA-PSK Encryption

CommView can display encrypted traffic from APs in real time, if you have the right keys. Under Options, check "Forced WEP decryption." Note that this does not mean CommView can crack encryption or derive the keys that are being used. It means that if you have and are authorized to use the keys, then you can read data as it appears on the screen rather than having to save them, save the packet buffers, and decrypt them later.

Just as there is no limit to what information goes out into that vast series of wires and satellites, routers and gateways that we call the Internet, there is virtually no limit to what anyone with the right equipment and software can intercept.

Conclusion

All things considered, CommView is one of the best and most useful applications available. True, it isn't cheap at $500, but it is well worth the price. It does what other programs costing several times the price do not do and, of course, you have the advantage of running in Radio Monitor Mode.

Spend a few hours learning CommView, monitoring, and working with the versatile filter combinations and you will realize what an excellent program CommView is.

ETHEREAL

Ethereal (http://www.ethereal.com) produces a packet sniffer, and what it does is display on the screen all or most—depending—of the information that goes past a particular point within a network. Whether it is wired or wireless.

If you know how to interpret this mass of information—and it isn't easy—you can see what people are doing on that network. In other words, you can see what Web sites they are visiting and even read their e-mail.

Ethereal is not as user-friendly as CommView; it is more complicated yet more detailed. It is more a diagnostic tool than CommView. And it does not convert captured data into graphics—meaning that what you intercept will not be the same as what someone else is seeing on their screen as they log on to Web sites or read their e-mail. Text, yes. Graphics, no.

At the top are the IP addresses. The source here, 192.168.xxx.xxx, is an internal address and the destination is the IP that this computer was monitoring. In the chapter on Internet Tools you can read about net.demon, with which you can look up these IP addresses and see what is there.

The bottom right corner of the screen is where the "plain English" text would be if this particular connection were to, for example, a Web site that was displaying a text file.

To the left is what the right characters look like in hexadecimal, which you should remember is a system of counting based on 16 instead of 10.

In the center between the horizontal bars is more technical info that you don't really need to use the program, but some of it will prove useful, depending on how far you want to go in learning about networking.

Spend some time with NetStumbler and CommView and learn the basics, read Help files and anything else you can find, especially about ports and protocols and filters. Then what you see with Ethereal will be easier to understand.

AUDITOR

While Knoppix (reviewed next) is a more general suite containing word processing, etc., Auditor is designed more for networking and wireless in particular.

Auditor is available for download at http://www.remote-exploit.org/. (However, as mentioned before, sources for downloads come and go, and new versions appear as they are created. It's always a good idea to Google for the latest versions and sources.)

The most recent version as of this writing is auditor-220604-01B.iso.zip. To repeat myself, this file is 518 megabytes and with DSL may take several hours to download. Yes, DSL is supposed to download at "up to" 144K, but the source may not make it available at that speed. If there are too many people trying to download at the same time, the server slows down. Try some of the different mirror sites till you find a fast one.

Once you have it, as with Knoppix, copy to another directory and unzip it there. Then burn to CD, remembering that this is an image.

Pop it into your CD drive, or maybe your DVD drive if the CD doesn't work.

When Auditor opens, you will be asked what screen resolution you want to use. Select one, and

Ethereal is more complicated yet more detailed than CommView.

123

it will start installing, which takes several minutes. Then, depending on the version, there will be icons to click to open programs or the FluxBox menu.

What you will see may be overwhelming—hundreds of individual programs that you can use. To try and provide even the basics of these tons of programs and utilities would fill a very large book and is way beyond what I can include here.

On the CD are Help and documentation files for you to read and learn from, including links to follow where Auditor users ask questions and provide answers.

KNOPPIX

The programs reviewed here so far all run on Windows although a few, such as Ethereal and NetStumbler, also work on other operating systems.

However, there are dozens of wireless networking tools that run on Linux and BSD but not on any version of Windows. These programs are usually better than most of what is available for Windows. More versatile, more powerful, and at very low cost or free.

These are, for the most part, tools used by advanced IT professionals, network managers, and corporate security personnel. This has always been a problem, because for people who use only Windows and have no experience with other operating systems, learning a new OS that isn't as easy as Windows takes some doing. And then there is the problem of installing it—either you would need a dedicated computer or would install it "dual boot" so the machine could run either OS. Dual booting is tricky and requires that you repartition the hard drive and reinstall Windows along with the new OS—Linux or BSD. And, of course, that means backing up all your files and reinstalling all of your programs. It can be a real hassle.

The good news is that there are suites, or collections of programs, that are burned to a self-booting CD that has the OS built in. This means you can reboot the computer with such a CD in the drive and have a running Linux computer without having to install Linux. This loads an operating system into RAM, from which the many individual programs run. So while it is running, it has control over the computer, meaning you cannot use any Windows programs. Once it is terminated, however, the Windows computer reboots and goes back to the way it was—nothing has been changed. Unless the operator does something he should not do.

Make backups of your important files just in case.

The first of these programs is Knoppix, which has two versions. One is 3.4 (the latest), and the other is called "Standard." Version 3.4 has KDE, a graphic desktop with icons for the programs, or suites of programs, of which there are many. It has games, Open Office, which is an excellent low-cost replacement for Microsoft Office, plus various graphic apps, multimedia for audio and video, and of course, wireless Internet tools.

Knoppix 3.4 can be downloaded from sites listed in Appendix A, but it is a good idea to Google it to find the latest version. It is easier to learn because of the KDE desktop with icons. The Standard version does not have KDE or program icons, but it does have a text menu called FluxBox. Both 3.4 and Standard are similar and contain most of the same programs. An important difference is the wireless applications that you will see when you use it. Both have Ethereal, which is much like the Windows version, and Airsnort for unscrambling WEP-encrypted traffic.

Before you decide on an application, be aware that some of these programs are not as easy to use, and especially to configure, as some Windows programs. The information here is sufficient to get you started, but you will need to do some reading of the Help files and make some command line changes. That is, you will have to open some configuration files with a text editor to change some of the settings. For example, if you want to change the rate at which Kismet scans the wireless channels, you use the text editor. Then the log files will show a record of the APs that Kismet has detected in a text file, rather than appearing on the screen as with CommView. But it is worth the extra work because Kismet will "see" APs that none of the Windows programs that I have tried were able to. I will come back to Kismet later.

Both versions of Knoppix are image files about 600 Mb in size and take two hours or so to transfer with ADSL. If you prefer, you can buy a working CD from the Knoppix site.

Sniffers and Other Wireless Applications

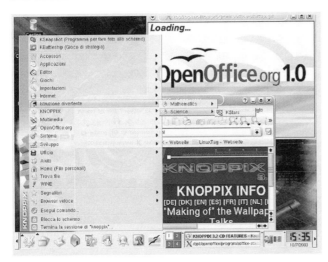

This is the opening screen, or desktop, for an earlier Knoppix version, 3.2.

Once the file is (finally!) downloaded, copy it to a directory of your choice, leaving the download where it is as a backup, just in case.

Depending on which version you get, it may be in any of several forms. If the file ends in .iso, it is ready to burn as an image. If it ends in .zip, you can uncompress it with Winzip, available at http://www.winfiles.com. Some other extensions may require a different application to uncompress them, such as .iso.bz2. You can use WinRAR, also available from http://www.winfiles.com.

Next, you need to burn the ISO Knoppix image to CD. If necessary, read the Help files for the CD burner you use so that you are familiar with making image files, rather than just copying, which will result in a CD that will not work. Remember: Image. ISO.

Once you have a working CD, reboot with it in the drive and wait for the opening screen. This may take several minutes so be patient. If you see only a blue screen with nothing else on it, the image probably hasn't been transferred correctly, but it is also possible, for a reason I do not know, that it will take much longer to get running. So, you might go out for a cup of coffee or something and give it 15 minutes. If still no go, then you need to make another CD. I had to try it three times before I got one that works.

Troubleshooting

If the CD was burned OK (you can use a file manager such as ZTree for Windows and see that files have, in fact, been burned to the CD) but it won't boot, you can try it in the DVD drive if you have one.

Another possible problem is that once it starts running, the cursor is stuck in the upper right corner of the screen and the mouse can't move it. This is what happened to me. If you are, as I was, using a standard PS-2 mouse, then hustle down to Circuit City and get a USB model. Kensington make a nice infrared wireless model, CC stocks it, and the salespeople will be glad to see you again.

With the Standard version you will first see the black cat that looks like Felix. Right click the mouse and you will see the FluxBox interface (menu) that lists all the programs on the CD.

Play with some of the programs if you like; this is a good way to learn to navigate through all that is there. Then look at your wireless card. Is the LED blinking? If you are using the Senao card, it probably is. Otherwise, maybe not, in which case you need to do some research on running another card, or get a Senao. They are, incidentally, available at Surf and Sip stores (see http://www.surfandsip.com for locations). Now, with Standard, right click anywhere on the screen to bring up the FluxBox menus, click on Wireless Tools, Kismet, then Kismet-Start. A window will open and, assuming you are within range of one or more APs, you will see lines of green and yellow text moving around. These are the SSID names, channel, etc., which flash on and off. Let it run for a while, and then have a look to see what it has captured.

To view the logs Kismet creates and to see a list of the APs it has detected, right click to get FluxBox again, and at the top under Apps click Root File Manager. At the top left you will see a window that says something like /ramdisk/home/knoppix.* Backspace and type: /ramdisk/var/log/kismet and hit enter. You will see a list of five or so files all starting with Kismet. Double click any of them and then view the files. You will need to become familiar with what is in each of the files.

Now, if you made a list of APs from another wireless sniffer, you can compare and probably will see some new entries. Using the same exact setup, on the Compaq (right) computer and grid antenna pointed in the same direction, there were nearly a dozen new SSIDs.

If running 3.4, try WaveMon (Wireless Device Monitoring Application) and you'll see a graph showing the signal strength of any AP you are associated with/connected to (the Senao card LED will be steady green) and able to get Internet access, and lots of other useful info. This is one of the few that show the transmitting power of the AP and the actual frequency. As you get more involved in wireless networking, this will become important.

If you use Knoppix for a while and start to become comfortable with it, you will realize that you can do just about everything you could with Windows. Almost. There are some programs that are not "ported" to Windows such as Web site building tools and graphics and drawing programs, but unless you need them, Linux could become your new operating system!

The Knoppix CD can be installed to your hard disk drive, which means it will run much faster and make it easier to save your log files, but be aware that you will lose any Windows programs and data.

To learn more about Kismet, see the FAQ here: http://www.kismetwireless.net/documentation.shtml.

About PCMCIA Cards

The Senao card (using a Prism chipset) worked automatically, without configuring it, with both 3.4 and Standard Knoppix. Neither one worked with the LinkSys PCI WMP-11 card (left computer), which is also Prism. Likewise the Proxim (Hermes chipset) card; neither of the two LEDs lit up at all. LinkSys WMP55AG (Atheros) also would not work. With the right drivers, perhaps all of these cards will operate under Knoppix, but I have not looked into this—I never intended this section to go beyond a brief mention of bootable CD operating systems. But I do tend to get carried away....

KISMET

There are several flavors of Kismet; all basically work the same, but the screens and menus are different. This one, gkismet, is on the Auditor CD.

Now, this next part can be confusing—it took me awhile to figure it out and I will describe how I did.

Start it and let it run till you see icons appear on the left side of the screen. You should see a little flag, like a pennant, and probably the icon of a tiny computer. To the left of that are the + and – boxes for expanding the listing.

The pennant represents an access point, and the other icon is called a "probe." Remember that

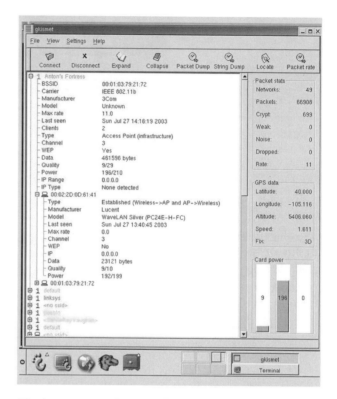

The icons you see here are from the Gnome GUI, sort of like Windows. Gnome is what this particular bootable CD was using. It could have been KDE or FluxBox.

in wireless we have two basic types of devices: APs and client wireless cards, either PCI or PCMCIA. APs have a name, an SSID, but client cards do not. A probe is a wireless card trying to connect to a specific AP, which is why you see an SSID.

To confirm this I used the left computer with the LinkSys PCI wireless card and tried to connect (associate) with an AP that was too weak and the attempt popped up in Kismet. I did the same thing again with a different AP and the same thing happened.

I will explain in more detail. Look at the pennant that reads Anton's Fortress at the top of the photo. This is an access point, as it says under Type. Below that is the other icon with the MAC of 00:02:2D:0D:61:41. Read what it says to the right of Type: Wireless –> AP and AP –> wireless. This means that Kismet picked up the signal from Anton's and determined that someone with the above MAC has established two-way communication with Anton's AP. Someone is using that AP for Internet access.

So what I did was use the LinkSys PCI card (left computer) to connect or associate with a certain AP and then looked at Kismet, which was running on the Compaq (right computer), to see if the MAC of the LinkSys PCI card I used was there. It was, and I also saw that Type showed that I was connected to this particular AP. Which, incidentally, was my own AP, the one in the Siemens router. In other words, in this experiment I was using only my own computers rather than connecting to an AP belonging to some unknown person or company.

Some of the other information you see in this screenshot is the same as you would see between the two horizontal bars with Ethereal, but here it is a little easier to understand.

Now, if you see the icon of the little computer, but it is not under a pennant—is not associated with any AP—then this is a probe looking for an AP to associate with. A wireless card like the Senao or LinkSys.

There may be many of these probes, especially in a large city. One of these is called Transit Surveillance Systems. This is a company in Anaheim, California, that makes security cameras for public transportation. When I first saw the SSID Transit Surveillance Systems probes, I Googled to find the listing for this company.

Now, I have not verified this, but what it appears to be is that Muni, the public transportation system in San Francisco, has security cameras on some of the buses that transmit on 2.4 GHz.

While running Kismet, I saw the Transit Surveillance Systems SSID pop up dozens of times in a given day. Since this is not an AP but does have an SSID, then it is a probe, and in this case it is, I concluded, the wireless client cards on the buses looking for their AP so that they could transmit the photos that were taken by and stored on the surveillance cameras.

Late at night when there are few buses running, I reset Kismet, and the Transit Surveillance Systems SSID no longer appeared.

WIRELESS CARDS

Several PCMCIA and PCI cards were used in wireless networking research.
Classic Orinoco Gold (B)
New Proxim card (B)
LinkSys WUSB 802.11b Adapter (B)
LinkSys WPC-11 (B)
LinkSys WMP-11 (B)
LinkSys WPC55AG (A, B, and G)
Senao (B)

Classic Orinoco Gold—It is no longer being made and so is becoming more difficult to get. If you can find one, snatch it up. This is an excellent 802.11b card. And it has an external antenna connection jack.

Proxim—This is a new card, from Proxim or Orinoco or Lucent or whoever is making it now.

Being a new card, it is understandable that there are some programs with which it is not compatible. It

The Proxim is a good card but has no external antenna connector.

does not work with the old version of NetStumbler, so you need the new 0.4 release. And it does not work with CommView, although new drivers may become available. It does work with some other wireless programs. And, of course, this will change by the time this is published, so check with the card manufacturer or software producer before you buy.

LinkSys WUSB 802.11b Adapter—Junk. Don't waste your money.

The LinkSys WPC-11 card—This is, in my opinion, another loser. It was difficult to get the drivers installed and difficult to get it to work at all. It does not have an external antenna connection, so it may work well at the local Internet café, but other than that, it isn't much good.

LinkSys WMP-11—This one surprised me in how well it works, notwithstanding that it was, and is, a little tricky to get working. When you reboot, sometimes you get an error message stating that some of the needed files were not installed. So you try to reinstall it and you get another error message stating that the files are already installed. If this happens, all you can do is use the Program Uninstall in Control Panel, then reinstall from scratch.

Now, as you can read elsewhere, I was able to associate, log on to, an AP and actually have Internet access using only the attached antenna after I turned the box around to face the open window.

LinkSys WPC55AG—This is a nice card. It started working automatically without installing the drivers, using those that Windows already installed. I traded some stuff for the card, which didn't have the installation CD, so for a while I just let it run as it was. Later I downloaded the drivers, and installation was painless. Very nice. I like this card as it captures all three bands. I used it in the field trip as described in that chapter, and the sensitivity was quite good. I definitely recommend it, unless you want and can find a similar (A, B, G) card with an external antenna connection.

Senao—This is, in the opinion of many users, myself included, the best overall B card available. The sensitivity is better, and the power output is higher, 200 mw compared to 30 or so for other cards, and this can be adjusted. Plus, it is easily put in "Stealth" RMM mode. CommView, for example, does this.

A LinkSys USB card. Not recommended.

Lest anyone think I am picking on LinkSys, this WPC55AG is an excellent dual band card. Unfortunately, it doesn't have an external antenna jack.

Installing the Senao is easy if you have the factory installation CD. If you do not, you will need to download the drivers and burn them to a CD, as trying to run them from the hard disk drive may not work very well. I had difficulty with this as I bought a used card without the CD.

Also, the Senao card works with Knoppix and Auditor; it took off as soon as I started them. Neither the Proxim nor the LinkSys WPC55AG did; it may be possible to get the right drivers, but I have not done so as of this writing.

Senao has a new A, B, G card but without the external antenna jack. Haven't tried it yet.

> HERE'S THE ANSWER TO THE QUESTION A FEW CHAPTERS back. First, I will repeat myself by saying that the reason I write things the way I do is to encourage the reader to think like a spy or a counter-measures technician.
>
> The clock in the photo appears to be a quartz type that would run on batteries for a long time. If this is true, then there is no reason for the power adapter seen in the photo. And it seems strange that no one would have noticed that it suddenly had a wire attached to it.
>
> Another possibility is that the clock was new, that someone had placed it on the desk after having built the transmitter into it. And maybe no one bothered to wonder how it got there.
>
> At any rate, the bug had constant power and it did the job. Until Mr. Hofmann found it.

Another card I haven't personally tried but was used in an experiment in the war driving chapters is the SMC-2532w-b made by SMC; according to the person I was working with, it is an excellent card. It is perhaps as good as the Senao, and it has an external antenna connector.

With most cards, setup is rather straightforward, except for some LinkSys models, which can be a real hassle, but one thing to keep in mind is that once you have installed drivers for one card, they may interfere with installing drivers of another. This is particularly true of CommView, which, as you have read, uses its own drivers. This is necessary to put the card in stealth mode. So, spend some time in Control Panel and Device Manager if you are running Windows. Drivers on the bootable Linux CDs are built in and work with most cards, but it is a good idea to read the Help files before you download and burn.

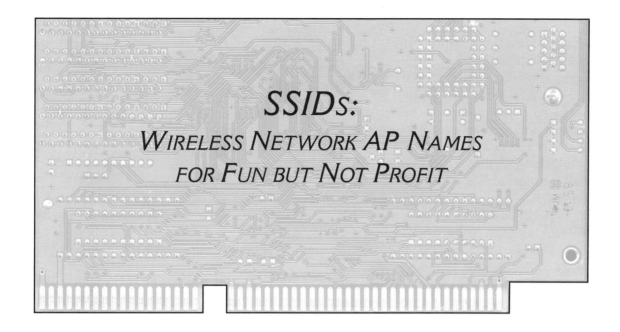

SSIDs:
Wireless Network AP Names for Fun but Not Profit

This chapter is about SSIDs, the names that people use for their wireless networks.

From my apartment, using different antennas, I have been able to detect about 190 APs and ad-hoc signals. All but a few are so weak that I cannot intercept packets—cannot capture any information from these sites other than to see the SSID register with NetStumbler and Kismet.

Some of them are very similar but not exactly the same, because in order to be separate from others within the same network, the name has to be unique. Otherwise, if you and someone else used the same exact SSID and someone wanted to connect to your AP but the other one was closer, they would get on that AP and not yours.

Citi-Net, for example, is a large network that provides wireless Internet to residents of a number of apartment buildings. They have a very sophisticated setup and broadcast on most of the 11 B channels. So they use names that are similar but not exactly the same.

As you start to build a list, you can't help but wonder about all these APs. Who owns them and what are they used for and where are they located?

Some of the SSIDs here seem like they should be fairly easy to guess. Hastings is probably the school of law half a dozen blocks from here, but I haven't verified this. And Lech Auto Air was a mystery—it would seem to be what the name implies, but I had not found a business with that name until the fourth annual WorldWide WarDriving competition, as explained in a coming chapter.

SSIDs such as Wireless, LinkSys, Belkin, SpeedStream, and Netgear are the defaults. When the owners set them up, they didn't make up their own name. So it is logical to believe that they are small home networks, since a business would change it to something other than default. Probably, but not necessarily.

Some appear to be the names of the individual nonbusiness owners such as Lefkowitz or Edrik/Christina or Steffan or cherye.

Apparently thrasher888 is about gaming, as may be Gamehenge.

Sfwireless.net is a free community AP for anyone who wants to use it. Justacrush is an "adult" site, part of, it seems, a large network. They are listed on Google.

Another mysterious signal is SST-PR-1. Do a Google search and you can read about the many ideas people have about what it actually is. Note that this is an ad-hoc signal, not infrastructure as from an AP. I happened to see SST-PR-1. So, just for the helluvit I punched it onto Google and it came back with lots of listings. People speculating that it is the United States Secret Service, which I kind of doubt (too obvious), and many who believe that it is from Sears vehicles—service trucks and whatever.

Now, as to the physical location of an access point, keeping in mind what you read about radio waves, finding them would take endless hours of walking around with a portable computer such as a Zaurus. War walking. If you like, do your own neighborhood survey, see what you come up with, and with some detective work maybe you will be able to figure out who owns the APs and what they do with them.

A BETTER WAY

If you have spent time searching Google or AltaVista, you know that while there are many sites that list "hot spots," these are all Internet cafés, coffee houses, hotels, and the like. But they don't list all of the other APs you have seen because they are not for public access.

But there are a few Web sites that list everything they have, regardless of what they are, assuming the listers even know what they are.

In the chapter on the Worldwide War Drive are links to the lists that have been compiled. And you just might find some of your own APs listed with their latitude and longitude. Plug that into a program like Microsoft's Streets & Trips and you just might find them!

After the fourth Worldwide War Drive stats were published, I looked up the SSIDs I have in my database and found a number of matches!

Meanwhile, I wonder about the SSID "Hot Girls come to Apartment 405."

Next, a review of a snazzy little gadget and then later the story about someone who found my AP. Meanwhile, here's a sample list of actual SSIDs I've found:

2WIRE835	Hyatt
ACTIONTEC	inmotion
BCJ Wireless	Jeff Network
belkin54g	Johnson
CarloNET	justacrush
Citi-Net Wireless	JustUS
cherye	krakle
citi-net	Lech Auto Air
citinet	Lefkowitz
citiscape	linksys
citinet-public	luckyspot
crepesoleil	METRO
DJOPEN	miketec.org
Edrik/Christina	musae
El_Gato (Infra)	Netgear
Fluxoz	NETGEAR
Gamehenge	Nick Tennison
Granite	popp net
GOM	Research
hastings	SFHome
HOME	sflan6
Hot Girls	sflan10

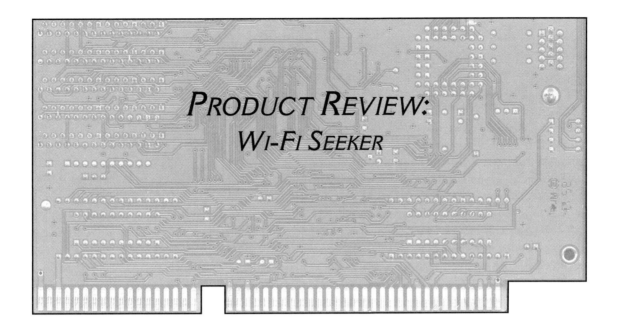

Product Review:
Wi-Fi Seeker

As the name implies, this is a device that alerts the user to the presence of wireless networking transmissions. The outside of the package says, "Inside this box is one very slick little piece of technology." The QuickLink Mobile Wi-Fi Seeker is a keychain wireless signal detector about the size of those car alarm remote switches.

BATTERIES INCLUDED

It works right out of the box—nothing to set or configure. Push the little blue button and the four LEDs start to flash. If a signal is detected, the flashing stops and one to four of the LEDs remain lighted.

So, being able to understand simple instructions fairly well, I pushed the lil' button and the lights flashed, then stopped, all four lit, picking up my home wireless network access point.

The enclosed instruction card says it works within 300 feet on a signal. I walked around my apartment, in the bath and kitchen and it read three LEDs. That's with two walls in between the Seeker and my AP. Outside on the street and halfway down the block I got one LED, and that's with the standard 2-dB antenna on the AP, lying sideways when it is supposed to be vertical, and through a brick wall. Not bad.

Meanwhile, being a hardware geek, naturally I had to take it apart. Three tiny Phillips screws. The printed circuit board is very well made; the lettering, the silk screening of the part numbers, is very consistent, which is one of the marks of a well-made product. It has a couple of tiny surface mount chips and a small silver tube, which I suspect is a crystal, and the usual other resistors and capacitors.

Did I mention batteries?

Wi-Fi Seeker comes with not one but two nickel-size hearing-aid types, which I suspect will power the unit for years. Something they might mention in their Web site.

FIELD TEST

Wi-Fi Seeker, it is claimed, will not be "fooled" by wireless telephones or microwave ovens. I don't have a microwave so I took it to the restaurant across the street and sat at the counter only 10 feet from the oven, waiting for Henry, the cook, to heat something. At that close distance Wi-Fi Seeker didn't register the oven at all. Zero LEDs.

Now, about distance.

One of the things about finding wireless hot spots is that you may not know their precise location; exactly where the antenna is and what is in between it and the outside world. So an exact measurement is difficult.

I got a reading of four LEDs while in front of the post office and then walked up Stuart Street what I guesstimate to be 300 feet, or half a block. The number of LEDs dropped from four to one, but since the display remained steady and didn't start scanning again, I was apparently getting the same AP.

All things considered, I like this "slick little piece of technology." It does what it is advertised to do and does it well.

Wi-Fi Seeker delivers on its promises.

The Wi-Fi Seeker wireless network detector was originally sold by Chrysalis Development, LLC, at http://www.chrysalisdev.com/. Last time I checked the link no longer worked, but the same device is available from http://www.smithmicro.com.

WIRED EQUIVALENT PRIVACY

THERE HAVE BEEN MANY ARTICLES, IN PRINT AND ON THE INTERNET, telling how insecure WEP (wired equivalent privacy) encryption is when used in wireless networking. They say that it is easy to break, and here and there it is implied that "everyone is doing it."

They make it sound—to the "average" non-tech person—like all you need to do is load up a new program, hit a few keys, and suddenly there it is, all of the wireless traffic, e-mail, and other confidential documents, right there on your screen in plain English. Yes, WEP can be defeated, but it isn't quite that easy.

First, you need to find an AP that uses WEP and that has a strong enough signal that you can capture and save the encrypted packets. Hundreds of thousands if not a couple million packets. Then, you need an application that will get and save these packets, preferably into (saved as) a single file or that the program can concatenate into one file.

Some applications can sort out and ignore traffic from other SSIDs and some cannot, but in any case the more "wrong" packets there are, the more intensive the operation. This is something to keep in mind when learning to use filters with an application that operates in true promiscuous mode.

Next, you need a program to do the decrypting. If you are running Linux or FreeBSD, there are a few applications available. If your OS is Windows, there wasn't much available except AirSnort, but to use it under Windows requires some tweaking. And there is Auditor, the bootable Linux CD, which will work.

I made an attempt by sending files from one of my computers to another and capturing packets with a third, the Compaq on the right. I was using AirSnort from Auditor but never was able to get it to work. I suppose eventually I could have but I just wasn't that interested.

So, unless someone is seriously determined to decrypt WEP on your network, you are probably safe. But just keep in mind that anything is possible.

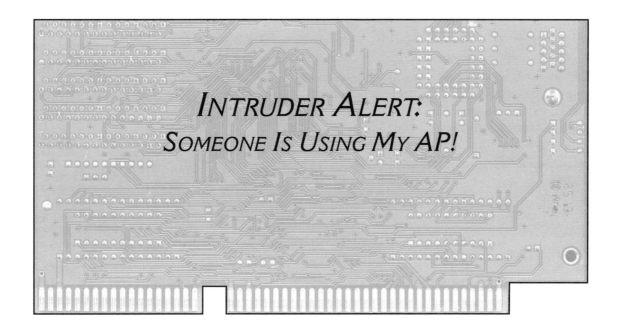

Intruder Alert:
Someone Is Using My AP!

I AM SITTING HERE WORKING ON SOMETHING OR OTHER AND I HAPPEN TO notice that the Wireless LED on the router is flashing, meaning traffic is going into or out of the device. Problem is, it shouldn't be.

So I double-check everything. The Sony (computer on the left in various pictures) is not powered on and this machine (center) isn't doing anything. That leaves the Compaq Presario portable on the right.

But the Compaq isn't doing anything either, so the AP is just idling. Only control signals, router broadcasting lookup tables, and such should be used and they won't cause the light to be on that much.

Someone has found my AP and is using it for Internet connection.

Fascinating! I have got to find out who it is and where they are. Not that I am concerned; it is a good learning exercise.

I will elaborate:

The router that the three machines are connected to has a configuration setup with which you can list the computers that are allowed to access the network and the Internet.

In other words, when the intruder connects to my AP, he cannot directly access any of my computers, cannot read or copy any files, but since I had "allowing Internet access" open, the intruder did, in fact, spend hours surfing the Internet through my AP.

I double-checked the configuration and once I had verified that I was secure, I opened CommView and scanned the channel my AP was on (9) until I had a list of everything that was operating within range of the AP antenna. Filters were set up so that only certain protocols would be displayed—HTTP and the e-mails including POP and SMTP. Then, nothing to do but sit back and wait.

So, I watch for a while and see the intruder as he connects to dozens of Web sites, mostly universities. North Carolina, Wisconsin . . . The intruder was reading files on the physical sciences, physics, and some chemistry, and then after a while he spent a couple hours searching through eBay.

SPOOFING THE MAC

Over the past several months I have been working on these articles, I have been entering SSIDs and MACs in a database. So, I note the MAC he is using and check my list, but it isn't there.

Next, I use the CommView feature that provides the manufacturer's name for the MAC.

Spoofed!

There is no listing for this MAC. But the list in CommView may not be complete (it is an added feature and not intended to be a complete database) so I Google it.

Again, no listing.

So now I know that the intruder knows how to spoof a MAC. This isn't someone who just happened to find my AP and use it without knowing it is not a free public access service—not an innocent kid who bought his first wireless card and was trying it out. Nope. I was dealing with someone who knows his stuff.

He stayed connected for several hours and I watched him go from one site to another, and from what these many Web sites were about, I began to form a picture. The intruder is a student, Asian, probably Chinese from Taiwan, has college-level knowledge of the physical sciences, has at some time in his life lived in Russia or has friends there and can speak that language to some extent.

A few hours later, he was gone and hasn't returned.

The Siemens router/access point. On top is the DSL modem.

I needed to change my setup configuration to work on a different project, and that required that I disconnect the AP, which meant also disconnecting the router. After I set it back as before, I never saw him again. I don't know if seeing my AP go down caused the intruder to believe he had been detected. Scared him off. I really wish I could have found him and learned where he was located—like I said, with the brick wall it couldn't have been very far away. I saved the log files and later when I had time, went through them, but couldn't find anything that would narrow down who my intruder was.

I was hoping like hell that he would do an e-mail check. Then I would have him. But alas, no. Anyone smart enough to spoof a MAC isn't gonna do something that dumb.

If he had stayed connected long enough, eventually he would have done something that would have revealed who he was. Now, since he was—had to be—so close, I could have taken a pocket computer or wireless PDA and done some war walking. But, alas, I don't have either. The Zaurus from Sharp, with a nifty Wi-Fi card, a nice little gadget that runs Linux, fell off my desk all of 2 feet onto a carpeted floor, and the backlight broke. Cost more to repair than it was worth. And toting the Compaq

> THE ARRANGEMENT I HAVE with my network configuration is rather inconvenient. If I want to scan for APs, some software will stop on my own AP and not scan any further. But since two of the three network computers work wireless through the AP, if I shut down the AP, they lose the Internet connection.
>
> There is no way to turn off the AP except by shutting the router down. Something I didn't know when I bought it. So, should you want a setup similar to this one, you might consider a separate router and AP. True, I could use CAT-5 cable but that involves changing settings, and it is easier to just do without one computer and use dial-up for the other. Meanwhile the main, center, computer is wired to the router so I just unplug the CAT-5 cable from it and connect directly to the DSL modem.

around isn't practical as it weighs too much—it's a full-size notebook computer.

The point of all this is that serious hunting and finding people who associate with an AP is no trivial matter. I will qualify that: If you have an AP located in a rural area and there isn't but one or two houses or office buildings or whatever anywhere close—within a few hundred yards or so—then you know where the intruder is located. In a large city with hundreds, thousands of APs operating, well, that complicates things. Remember that radio signals–including Wi-Fi transmissions—are unpredictable and can bounce off buildings and be detected in places that are not in the direct signal path. So finding an intruder is no trivial matter.

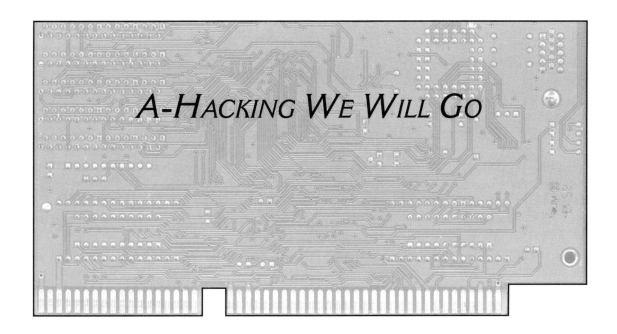

A-Hacking We Will Go

WITH WHAT YOU ALREADY KNOW ABOUT COMPUTERS IN GENERAL AND WHAT you have read so far, you should have a good understanding of wireless networking. And if you have obtained and learned to use some of the programs reviewed here, including the Auditor self-booting Linux CD or the "Frisbee" Free BSD disk, you know how to search for wireless APs and connect to them (associate) so that you can use them to get a free Internet connection. Just keep in mind that while it may not be unlawful to detect APs, it is against the law to connect and use them unless they are known to be free to the public.

It is also, sometimes, possible to take control of the network. To demonstrate this, Evil, a friend from IRC (an Internet Relay Chat channel #SF2600) came over. He set up his portable computer, a Sharp model that runs Free BSD, and an SMC card connected to the big antenna.

Incidentally, it was interesting to find out that with this combination, he was able to detect the same SSIDs as I did, which allayed my fears that there were many wireless APs that I could not pick up.

After a careful scan, with Evil turning the antenna around 360 degrees and making a log of what is out there, it was time to get to the business of serious hacking.

For a target, we started with my AP. First, it was detected using a sniffer, and from the log file, details were obtained. Evil then tried to connect to get access to both the Internet and the computers on my small network.

He was unable to do so, even though I use DHCP (Dynamic Host Configuration Protocol), because I programmed my router to allow access only to my own machines.

This was not by any means a dead end, as you will see later.

Next, we selected an AP that clearly stated it was for free public access (I will not identify it other than what you will read here). Now, this particular AP is part of a network—it is set up for multiple users and so uses a router, and that router uses DHCP, which I will explain in more detail elsewhere in this work. I will also repeat some things I have mentioned that will put it all together.

Every computer on the Internet has to have its own unique identifier, just as do telephones; otherwise, there would be no way to connect directly to them. Now, suppose you have a number of computers and you want each of them to have Internet access. To do so you would need each one to have its own IP and own account and therefore you would be paying for all these accounts. Not a good idea. So what the IT people do is assign each computer a different IP through a process called subnetting.

In order to pass the CompTIA Network Plus certification test, I had to learn how this works and be able to manually, with paper and pencil, actually calculate the IP for each of a number of individual computers, and also figure the maximum number of machines that can be used. Without getting more complicated than necessary, this involves taking one IP in its dotted quad notation (219.123.23.117, for example) and converting each part to binary, then "borrowing" some bits from one of the four quads (depending on the class of the IP) and using them on another quad. With DHCP this is done automatically through the programming in the router.

So, when you walk into a wireless café, the AP detects the probe signal from your wireless card and the router assigns you a temporary IP so the connection can be made. Now you can have a cup of coffee and an overpriced lemon bar, check e-mail, and read *The Wizard of Id*.

PROGRAMMING THE ROUTER

The Siemens router I have uses a Web browser to make and change the settings, including whether or not DHCP is used and various other things. (It is also possible, with most routers, to make the settings manually. Here, you would enter the computers that are permitted access to the network and/or the Internet.) So from the main (the middle) computer, I type the IP, the address of the router, which is 10.0.0.10, into the location line in Opera, and after the user name and password, I get the setup screen.

Now, if I wanted to, I could use either of the other two computers (the left one, which incidentally uses 10.0.0.14, or the right one, which is 10.0.0.13) to access the router. I believe I know what you are thinking. If we can associate with an AP and find the IP their router uses, and get past their password assuming they even use one, then we could control the router, right?

Yep. Most definitely.

In the first example, using my network, Evil didn't attempt to find the IP that the router uses and even if he knew it, there is still the administrator's user name and password. This is much like a burglar who attempts to get into a home only to find strong locks and an alarm system and moves on, looking for a place that has neither. And in the case of wireless networks, there are plenty of them.

So next, we tried the public access AP. Once associated we had the IP of the gateway—the router—and the brand name. Now, where I use the IP 10.0.0.10 for access to my router configuration utility, this one used 192.168, which you may recall is a block of IPs reserved for internal use.

Many APs use the default 192.168.0.1, which we tried, and indeed it worked. The next step would be to get past the password, and we got lucky—whoever set it up used the default, which is not that unusual. So, we were able to get in the router and make all the changes we wanted. We could have:

Evil prepares to go hunting for APs.

> THESE BLOCKS OF IP NUMBERS are reserved for private internal network settings. In other words they are used within a network:
> 10.0.0.0–10.255.255.255
> 172.16.0.0–172.31.255.255
> 192.168.0.0–192.168.255.255
>
> It works like this: If you set up a LAN and want every computer on that network (called a host) to have Internet access, without subnetting, each machine would be required to have its own individual Internet DSL account. Obviously not cost effective.
>
> Now remember that every computer has to have its own unique number, its IP. So we use subnetting to provide each with its own IP, and those IPs used are within any of these three ranges.

- Re-routed all e-mail to the Sharp computer here at my apartment, copied it, and decided whether to let it arrive at its intended destination. We did not.
- Changed DHCP to manual subnetting and controlled who would be able to use this AP and who would be blocked. If we knew who someone using this AP was from his MAC, we could have arranged to block his access and for him to see a message stating that he was no longer welcome because he spends too much time watching "college girls take it all off." We did not.

There are other things we could have done.

We could have gone to a place that has free wireless access and spoofed their AP so that others there would be logging on to, connecting to, our portable computer without knowing it.

We could have placed phony ads such as "Free lemon bars with coffee purchase" or advised people that it has just been discovered that the latest coffee bean shipment is contaminated. We could have obtained the login info ad password of anyone who connected. And it would not have been difficult. But we did not. This was an exercise, a demonstration of how easy it can be to take control of some wireless networks.

Could we have been caught breaking into someone's network?

First, how do we define "breaking"? This AP is open, available to anyone who wants to connect. For free. So, by accessing it, associating, or connecting to it, we didn't "break" into anything.

As to accessing the router setup menu, what we did was type different IPs into the browser window. This is what anyone would do to log on to any Web site except that we used the dotted quad notation (192.168.x.x) instead of the name. We observed what we saw. We looked through the menu selections to see what was there, but again, we didn't change anything.

Now, as to being caught, we might have been if the people who own this network had been running the right software. As to them finding out who we were, this is very unlikely unless we did something stupid, such as sending e-mail through their server using one of our real e-mail addresses, or logging on to our own Web sites, or accessing an FTP site where we required and used a login name and password—all of which could be traced back to us.

And where we were geographically (my apartment) this is even less likely, as you read in the Intruder chapter. If they even noticed that we were into their router configuration, they would have to take a portable computer like the Zaurus and try war walking to find us. And again, as you have read, radio waves do strange things and aren't that predictable, so where would they even start? And what would they be looking for? The MAC of the Sharp computer? Hell, we can spoof that whenever we want.

So far, we were keeping a fairly low profile. But what if we attempted to take control of an AP where the password was not the default for the brand being used?

Enter some utilities that run on the BSD Evil's computer uses—Nmap, Ettercap, and Airsnarf. Running them would give us what we need to take control of the network that the owners, having it password protected, thought was safe.

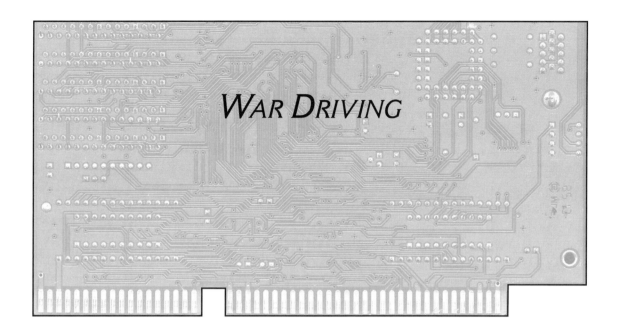

WAR DRIVING

THE TERM "WAR DRIVING" PRESUMABLY COMES FROM THAT REALLY DUMB movie *WarGames* (1983), in which was described the old technique of using an automatic phone dialer to find phone numbers for computers. Back in the days when dial-up modems were the only way to get online (before DSL and ISDN) and when the Internet was limited to a select few, these dialers would find RBBSs, remote bulletin board systems that people set up to communicate with each other and share files, and also business computers that used the same modems. And back then, security had not even been heard of, for all practical purposes.

So it was easy to cause a simple buffer overflow by typing in a long string of characters, which dropped the operating system to the command prompt. Root. Just the same as if you were at the keyboard of that machine. In other words, you could edit copy and delete files, and build a user name and password that the system operator ("Sysop") wouldn't necessarily know about.

Well, that was all done from desktop computers through the phone lines, as there weren't any wireless computers.

So far, you have read a quite a bit about wireless networks and have a good understanding of how they work and how they can be monitored from a fixed location. So now, let us look at some of the vulnerabilities of wireless systems and how they can be monitored by people who go around looking for them.

War driving is the technique of using a portable computer, a notebook, or even a wireless PDA such as a Zaurus (which runs Linux) with a wireless card and a small antenna to find APs. A fairly sophisticated setup would include a global positioning system receiver that would log the actual location, latitude and longitude, where the AP was detected. And all this can be done automatically while just driving around. Or walking. While just sauntering around in the business district of any city, which naturally is called war walking. Everything, including GPS, is easily concealed inside a backpack or in an attaché case.

You could even obtain a messenger bag, which along with a clipboard and handheld radio will get you into practically any office building. Messengers, like the

THE FOURTH WORLDWIDE WARDRIVE

THE BIG EVENT IN WAR DRIVING took place in June 2004 with hundreds of teams searching and logging access points all over the globe. This demonstrates not only how popular war driving is and how so many people all over the country can work together, but it also produces a massive database of APs. Within the United States, several hundred thousand were detected and records were uploaded to http://www.wigle.net.

Thanks to this huge list on http://worldwidewardrive.org/ (which includes the latitude and longitude of the AP) I was able to find a few of the APs on my list from the chapter on SSID.

The antenna in the photo looks familiar, no? Unfortunately, I could not get away to take part, so I loaned it to someone who could.

Above: A well-equipped geekmobile.

Left: Set up and monitoring outside Reno, Nevada.

mail delivery people, are "invisible"; no one pays any attention to them.

Depending on your equipment, the APs and what—if any—security is being used, and keeping in mind what you read about radio waves and antennas, you might capture and store long lists of APs with their SSID, the MAC, e-mail sent by anyone from the receptionist to the CEO, passwords, and log files showing the Web sites that the employees are connecting to.

A story that has been told many, many times on various Internet sites and IRC channels is about how the IT person maintaining the company network can see who is doing what and catches management personnel visiting "adult" porno sites. And logging it. Without using the word "blackmail," it has been suggested that this knowledge contributes heavily toward job security.

Is war driving legal?

Yes.

War driving is not a crime. A recent FBI e-mail stated it this way, however:

"Identifying the presence of a wireless network may not be a criminal violation; however, there may be criminal violations if the network is actually accessed including theft of services, interception of communications, misuse of computing resources, up to and including violations of the Federal Computer Fraud and Abuse Statute, Theft of Trade Secrets, and other federal violations."

It is, however, each individual's responsibility to ensure that they do not violate any local, state, or federal laws that may pertain to their area.

This is quoted from the http://worldwidewardrive.org/ site.

PREVENTION: WHAT CAN YOU DO?

If you happen to be an IT employee of a large corporation whose job it is to secure communications, protect against hackers, war drivers, and anything else that could jeopardize the company network, then you have probably attended workshops and lectures and (endless) meetings where you have learned the very latest techniques in combating "cyberterrorists."

Your company has likely spent tons of money on hardware and software and (Can you say "Windows"?) patches to secure your data, but you still have problems.

The article about bank robbers and cyberterrorists should be enough to make you see that there is no way to have total security.

So what can you do?

A wireless PDA like this Zaurus can travel with the war walker or driver.

Marks such as these on a building can indicate there are APs that have signals that can be intercepted.

Read this work carefully. There is a lot of useful information on security in the text but not all in the same place. Apply what you learn, teach your employees Common Sense 101, and hope for the best.

Arrange for a site survey as described in the next chapter, but—and this is a biggie so I will repeat myself—it isn't enough to hire a counter-measures team just to look for "rogue" (unauthorized) APs. You need to know how far away these APs (as well as the PCMCIA cards) can be monitored. Carefully done surveys measuring signal strength from different locations and a detailed report can make this clear.

Then you can rearrange, relocate your APs, place them so that something like file cabinets are between the AP and the outside world, or even purchase shielding material that can be placed as needed.

A few years ago I read a story (which may or may not be true) about how the feds were using cell phones and making calls where they claimed to have lost something very valuable, like a diamond ring, in a particular area. Then they stake it out, looking for people who were unlawfully monitoring cellular and went looking for the gem.

What with all the war driving going on, I suppose it is possible that one law enforcement agency or other might set up APs intended to entrap people. A "honey pot" like on the Internet, something like that. Maybe an AP with an SSID that anyone using NetStumbler might see and want to investigate. Like I have written about.

If I see an SSID like "Hot Girls" (see the chapter on SSIDs) and try, for obvious reasons, to connect and unless I have my MAC spoofed, this could eventually lead back to me.

WAR CHALKING

Once upon a time, so the story goes, back in the 1930s, the Depression years when hobos wandered the country, there was a system of marks that hobos would leave to show places of interest. These marks would indicate a house where the residents were generous enough to give away a little food or, to quote Roger Miller, "... Every lock that ain't locked/When no one's around."

Today, war walkers supposedly leave chalk marks on sidewalks or on the sides of building to show where there are APs, wireless networks that have signals that can be intercepted, data that can be captured.

Well, naturally, the media made a big deal about it a few years back, but it really isn't. For one thing, there just aren't that many people walking the streets with a Zaurus in their shirt pocket looking for APs, and anyone who wants to chart or survey a particular area can do it more efficiently by war driving. There's more room for equipment and antennas, it is faster, and with the GPS I mentioned, a large area can be mapped and the results uploaded to a Web site, of which there are quite a few.

Off and on over the last five years, I have worked as a walking messenger (that's how I know that we delivery people get into most office buildings without being noticed), and in that time, I have never seen a single war-chalking symbol.

WAR WALKING: A FIELD TRIP

SHORTLY BEFORE COMPLETING these chapters on wireless networking, I made the rounds of a few of the many wireless cafés and other hot spots within a few miles of here. For this experiment I used the computer on the right, the Compaq Presario notebook, a Pentium III with NetStumbler and Capsa. Only one card was used: the LinkSys WPC55AG, which receives all three bands, A, B, and G. The reasoning behind this is that, first, within such an establishment the signal will be strong enough that the card will capture packets as well as eavesdrop on other wireless customers, and second, that this is the way most people would be doing it. In other words, without an external antenna. I wanted to see what they see.

First stop was a restaurant where I sometimes eat, which is next door to a Wi-Fi café, the Crepe Soleil, located three blocks away on Polk Street. Incidentally, as I have already mentioned, it is interesting that with the Senao card and grid antenna I can pick up their AP even though there are four- and five-story buildings between here and there. It is so weak that I cannot intercept any packets, but it is there. Perhaps an example of multipath? I have not found any other cafés of the same name.

So when I got there and set up, I verified that the MAC is the same. Radio waves are unpredictable. The signal was strong as shown with NetStumbler, but Capsa was not capturing any packets and I was unable to get a connection.

So, after aspirating a chef's salad, I went in to Soleil to ask about this.

There were two people behind the counter, but neither of them knew anything about it. They did not know who was providing the network connection or their IP or SSID; nothing.

No one there was using a Wi-Fi card, so I split.

Then, I meandered down Polk to the Quetzal and, upon asking, was told that, yes, they have wireless access in addition to a dozen wired terminals, but that "Wireless isn't working today." A person could get discouraged.

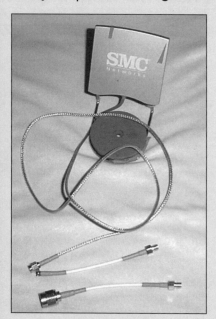

Here is a little SMC external antenna, used at wireless restaurants, and the pigtails that are included. Very nice.

Next was a small café on Post Street, one block to the east of my apartment. (I cannot detect this AP at all from home.) They have a sign out front advertising Internet access and have two desktop PCs. Beside each is a laminated plastic sign with their rates, like $2 for 15 minutes or $7 per hour.

I bought a bottle of imitation fruit juice and settled into a table in the back and unpacked the Compaq. No sooner than I had it fired up, I was connected to their AP. Very strong signal, said NetStumbler. I opened my browser, Opera, and as it is set to continue from the last session, Google instantly popped up.

I wonder if the owners know that anyone with a Wi-Fi card can use their network for free, while those on the terminals have to pay.

Enough for one day, so I went home.

Next day I went downtown, to Rincon Center for a salad, a root beer, and a free "dinner roll" that could have doubled as a hockey puck. There were, said NetStumbler, five APs; two were encrypted and the other three were fairly weak. I was not able to connect to any of them and none of them had any traffic that CommView was able to detect.

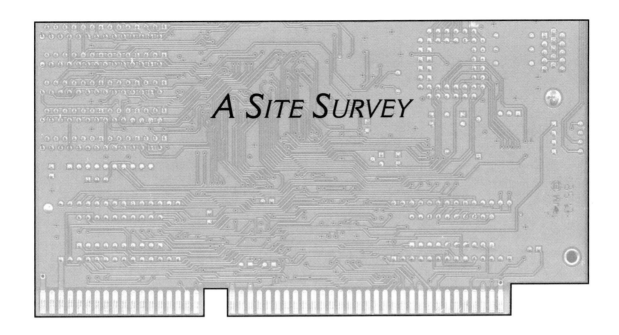

A Site Survey

When a corporation or even a small business decides to go wireless, it may have a survey conducted to determine the best places for the APs so that coverage is over the complete facility. There are people, experts, who do this and so the new network is installed and everything is fine.

But we live in an age where electronic spying is common, especially industrial espionage where one corporation spies on another. So periodically they have TSCM electronic countermeasures teams come in to sweep the premises for listening and video devices, and for rogue APs that an outside spy from the competition or an employee who has been bribed might have connected to their wireless network.

But according to my research, which sometimes involved calling some of these experts and pretending (social engineering) to be a potential client, none of the crews that set up wireless networks and none of the electronic countermeasures teams check to see how far away their APs can be monitored. And yet it isn't a difficult process at all, as you now know.

A site survey as I would conduct it would go something like this:

Objective—Determine at what locations and distances the signal from the AP can be detected, and whether or not the signal is strong enough for an unauthorized person to:

- Monitor the AP and copy the data flowing through it
- Access or associate with the AP; use the AP for Internet surfing.
- Take control of the AP by accessing the router. (More on this coming up.)

Methodology—Having consulted with the client, a list of all SSIDs (station set identifications, also known as service set identifiers) would be obtained, along with network names and Media Access Control (MAC) codes.

Monitoring equipment would then be set up at various locations around the premises, and measurements made.

- Equipment Used—Compaq Presario notebook computer running Win 2K, Linux, and street-mapping software
- Senao, LinkSys WPC-11, and Lucent/Orinoco PCMCIA Wi-Fi cards
- Windows NetStumbler wireless sniffer and CommView wireless packet capture and analyzer
- Linux Kismet
- Antenex 11 dB Yagi tripod-mounted antenna
- Grid 24 dB tripod mounted dish antenna

Next, the report.

(The following report is, of course, fictional, based upon the old radio program *Lum and Abner*, who operated the Jot 'Em Down Store and Library in the little community of Pine Ridge, Arkansas. I could not have used a real report unless I sanitized it to the extent that it would be filled with blank lines. And while it is also simplified—a rural community with only a few hundred residents—it does reflect how a professional wireless survey would be done. The principles, methodology, and equipment are the same.)

THE SURVEY REPORT

Mr. Lum Edderds
Jot 'Em Down Store and Library
Pine Ridge, AK 71966
Phone: Two long and two short rings

Dear Mr. Edderds,

As requested and per our written agreement, a discreet site survey of your wireless network was conducted as follows:

- Sweep for all access points, both authorized per your list and "rogue," or unauthorized.
- Determination of the locations at which both authorized and rogue APs can be intercepted.
- Plotting of latitude and longitude of these locations using mapping software.
- Search for "probe" wireless client cards within range of all APs.
- Physical inspection of your corporate offices.

The first phase was instituted at 20:00 hours on 13 April and consisted of concealing a tripod-mounted 24-dB magnesium grid antenna inside your new Jot 'Em Down Library bookmobile. This high sensitivity antenna was connected to a Compaq Presario notebook computer using a Senao NL-2511-CD wireless PCMCIA client adapter. Several software applications were used as described later in this document.

Immediately on starting Network Stumbler, we could see five identifiable signals, and after consulting the list of APs provided by your IT manager, and eliminating their respective MAC addresses, there was one remaining transmission that was not authorized, emanating from the direction of your corporate offices.

A rogue AP was discovered.

On the basis of that rogue signal, we began a

The Garmin GPS-72 is difficult to use, unnecessarily complicated, and unpredictable as hell. The buttons do not always do what they are supposed to, and it shuts off unpredictably even with fully charged batteries. It also requires a proprietary cable for computer interface, which is way overpriced. Several e-mails to Garmin were not answered.

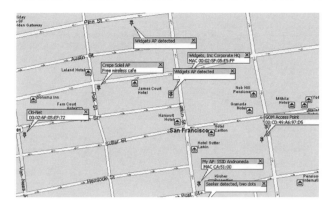

A map like this can be generated to show the APs in a given area.

360-degree sweep using CommView Wireless and detected a computer with an active network card. Switching to the Linux-based program Kismet, we were able to determine that the wireless card with that MAC address (44-23-EF-F0-B3-2D) was monitoring your network, including the one rogue AP.

By moving the bookmobile through the immediate area and activating both our Global Positioning System receiver and associated mapping software, we were able to establish the precise location of this computer. Next, a Zaurus 5000 PDA with D-Link wireless card was hand-carried around the building in which the intruder was monitoring your network in order to eliminate possible multipath signals.

It was concluded that this computer was within the structure, which is the home of Mr. Squire Skimp.

Since Squire was using an application that did not put his card into radio monitor mode, we were able to capture his DNS requests and then using the Zaurus handheld computer, zero in on his home. So, there is no doubt that Squire was monitoring your wireless traffic. And it is probable that he, or someone in his employ, was responsible for placing the rogue AP.

We then began a physical inspection of your facility. This rogue AP was hardwired to your router, and the CAT-5 cable was hidden inside an old unused garden hose, leading through a knothole in the pine paneling and placed on a windowsill.

Recommendations—As with any situation where positive surveillance has been confirmed, we meet with the client and outline the options available to them.

1. Leave the rogue AP in place and transmit information that is likely to cause the persons monitoring your network to do something that would cause them to be caught in the act.
2. Have your IT people reprogram your router. Disable DHCP and manually set who can access the LAN. Obtain and install a switch or a switching router to segment.
3. Redesign your network to reduce the number of APs being used and locate APs so that they are not placed in front of windows.

In a real survey for a real company, I would have included a comprehensive report including one or more maps showing the precise location—within 50 feet or so—where signals from a given AP were received, the latitude and longitude, and in some cases the actual street address.

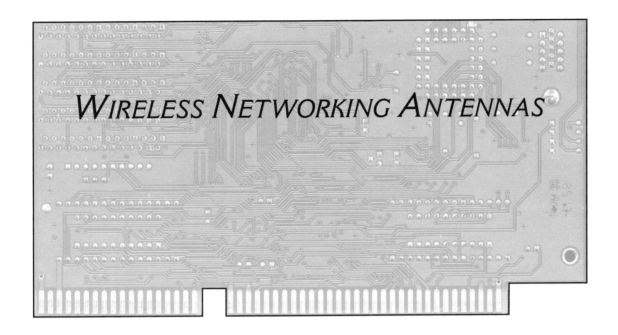

Wireless Networking Antennas

A RADIO WITHOUT A DECENT ANTENNA IS LITTLE MORE THAN USELESS depending on, among other things, the type of transmissions listened to. FM broadcast requires little more than a small telescoping-type antenna or the kind built in to portable radios because the power of the stations is so high—tens of thousands of watts in ERP (effective radiated power), which refers to how the signal is "amplified" or actually concentrated using directional or "gain" antennas.

With Wi-Fi cards, which are actually little two-way radios, this is nonetheless true, even more so. Use the built-in on a typical card and you might get a couple hundred feet, maybe more depending on conditions but maybe much less. Remember that radio waves do strange and unpredictable things.

Using something like the 12-dB Yagi pictured in the Intro to Wi-Fi chapter, it is possible to establish solid, reliable two-way communications over much greater distances, possibly several miles, if there are no obstructions in the signal path.

In Sweden a link was established from a balloon at a distance of about 185 miles. Now this was using much more power (six watts) than any Wi-Fi card is capable of (they used an amplifier), but the point is that distance is virtually unlimited given the right equipment and conditions.

Here in San Francisco, links have been established across the Bay to Berkeley, and someone somewhere claims a link of 75 miles (unverified) using an ordinary Orinoco Gold card.

To put this in better perspective, consider that licensed ham operators have been having long-distance two-way radio communications for decades, using low power, but with high-quality antennas that are matched to the equipment.

One warning: Before you decide to set up a card, such as the Senao, with a high-gain antenna, and especially if you want to use an amplifier, remember that you will be transmitting a very strong signal that could interfere with other wireless installations, such as those used in hospitals. And also, the use of amplifiers is not permitted in the 2.4 GHz band. If you want to set up a link between yourself and another person, observe what is in the signal path so as to not cause harmful interference, use the most directional antenna available, and if your card can have the power level adjusted, use no more than you actually need.

BUILDING WI-FI ANTENNAS

If you like to build things, then you may want to try to make a wireless network antenna. A good place to start is the Web site of Seattle Wireless, listed in the sources section, where there are a number of articles on how-to. Now at first, this may look like an easy project, as it might be, but again, please read through the articles, especially the list of parts, materials, and tools you will need. It could be more expensive than you anticipated, and also it may mean going back to the hardware or plumbing supply place again and again. Do you already have a hacksaw, a tube cutter, X-Acto knives, sandpaper, an electric drill, a good set of files?

I decided to build a 24-dB Yagi, which required PVC and copper tubes, nuts, bolts, washers, threaded rod, something for the backplane (I used a stainless-steel salad bowl), and miscellaneous other small hardware, plus a way to mount it to a tripod.

It took a number of trips back and forth to several hardware stores (none of them had everything I needed), and then I discovered that the washers didn't all fit on the threaded rod, so I had to use a rat-tail file to enlarge the holes. Then, I was unable to find the right PVC fittings so had to do a lot of cutting and sanding and drilling.

Finally I gave up—it was taking too much time—and bought an 18-dB flat panel type, and later the 24-dB grid. A better way for a higher-gain antenna is to use a parabolic reflector. What I didn't know at the time is that the Direct TV types, which are on 5 GHz, will work, according to one of the geeks on the local IRC channel. So, have at it if you like, but it doesn't hurt to know what you are getting into and, most of all, determine that everything you need is available before you start.

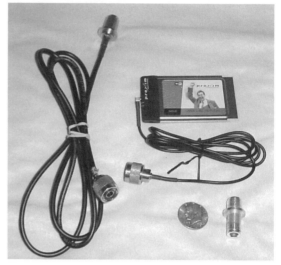

A 6-foot pigtail on the Proxim card. Below, an "N" female adapter. Should you buy a pigtail cable and antenna that both use a male "N," this adapter can be used to connect them together.

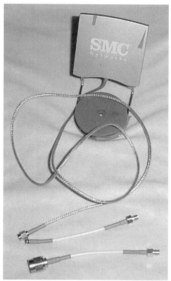

The SMC 6-dB antenna uses an SMA connector. Below that is the pigtail with which it can be used with the Senao card. The bottom pigtail is an MMX to TNC.

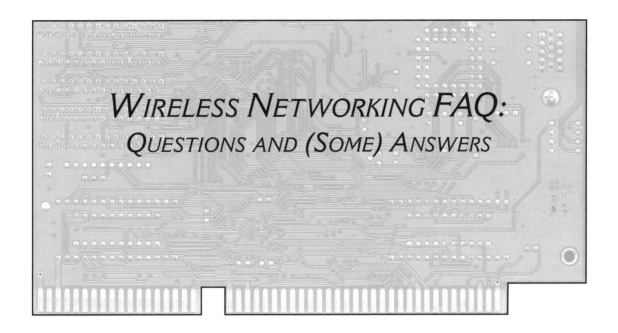

Wireless Networking FAQ:
Questions and (Some) Answers

Much of this information may be found in the text, but since networking is a tad complex, it seems like a good idea to put some of the answers and unanswered questions in one place.

This is for both Windows and Linux. Originally I intended only to mention that bootable CDs running Linux were available and not get into details. Then I discovered Auditor and, along with that, the many applications for wireless security and analysis it includes. But remember that this is Linux, not Windows, so it will take some time to learn.

Q: Why does Kismet (Linux) find some APs that NetStumbler or other Win applications do not even with the same card?

A: Some APs do not broadcast their SSID. So when NetStumbler, for example, sends out its beacon, these APs ignore it and do not respond—do not send a return signal. So NetStumbler does not see them.

Kismet monitors all traffic and sees all signals regardless of whether the AP has its SSID suppressed. So you may see—depending on the application—something like <no ssid> in Kismet or just an empty space with other sniffers.

Q: I see SSIDs like ^|^t/vG^H^W. What the heck is this?

A: I don't know. Apparently Windows XP somehow generates random garbage characters that end up getting into the system, the AP, and making these changes in what the SSID is supposed to be.

Q: I see one of these "garbage" SSIDs, but then it changes into what looks like a legitimate SSID such as DEFAULT and then to JoesPlace and whatever else. But the MAC remains the same. What's the deal?

A: I have no idea and no one I have asked does either. Usually an SSID is set manually; maybe someone just sits there and changes it whenever they want for whatever reason, or perhaps someone has come up with a utility that does this at random times.

Q: I see an SSID Transit Surveillance Systems now and then, and if I let Kismet run for a long time, I see more and more of them but they all have different MACs.

A: A Web search shows this to be a company in Anaheim that makes digital cameras for public transportation vehicles. The signals I see are all Probe Request (searching client), which is a wireless card looking for an AP. As to why the quantity of these same SSIDs (but with unique MACs, remember) increases, it is, I reason, the increased number of buses within range of my antenna.

Q: How do I understand all this stuff about different modes for Wi-Fi cards?

A: By reading everything you can find including the documentation files for any program you are using. The one for CommView is excellent as are those for the Auditor CD.

However, it isn't necessary to know this in detail any more than it is to understand all of the information you see revealed with a packet sniffer. It mostly comes down to this:

When you set up a Wi-Fi card (PCI or PCMCIA), you usually need to install the drivers for that specific card. However, some programs have their own drivers. CommView, for one, uses only its own driver that puts the card in "stealth" mode.

What the card does, then, depends on the drivers and/or the application being used.

For example, I plugged in the LinkSys A, B, and G card and fired up CommView and it worked. On all three bands.

Then I installed the software that comes with it and got the main screen where I can set it in either ad-hoc and select the channel, or in infrastructure mode where it will scan all channels looking for an AP. So, using the LinkSys drivers, you are limited in what you can do.

You have two modes: infrastructure, where the card is looking for APs, and ad-hoc, where you set the card to connect directly to another.

With some card/driver combinations, such as the ones that come with the Senao card, you automatically connect to (associate with) the strongest signal the card detects. This is why you log on to the APs at Internet cafés when you get close to them.

Or you can scan, look for SSID names, and then make a profile for that AP that will override this and connect to the one you want, assuming the signal is strong enough. This is called managed mode.

Q: How do I understand all the stuff I see using a wireless sniffer?

A: Like the above question, unless you intend to become an expert, there is no need to. If so, be prepared to do one helluva lot of reading and experimenting as well as spending a lot of cash for different types and brands of equipment.

Using the programs reviewed here, you will be able to make an accurate survey of a given area and report this to the client who has hired you, detailing any rogue APs that might be on their network and the location and distance at which it is possible to associate with their system and intercept data.

Q: The numbers indicating how strong a signal is seem confusing. They are different for every application I use. How can I understand all this?

A: They are confusing. NetStumbler lists them several ways, by relative signal strength, signal to noise ratio, etc., while other applications list by dB level. What counts is if you can or cannot get them strong enough to intercept packets. Consider using the numbers only for reference.

And as above, unless you intend to get deeply involved in the technical details, it really isn't necessary. You can make notes of the readings from a number of APs using different programs and compare them and see what in one program equals in another, and after a few examples it becomes clearer. But again, if the AP is strong enough for your purposes, that's what counts.

With NetStumbler it is the SNR (signal-to-noise ratio) that is the most important. More than likely only a few of the APs it sees will have numbers high enough that you can connect to. Anything below about 13 is unlikely. My AP sitting here next to the computer running Stumbler reads average 55, a solid connection. Anything in between is iffy.

Q: Will I ever be able to identify all the APs I see?

A: If you live in a rural area where there are only a few, maybe. There is a better chance in such an area that people will know their neighbors and the local businesses. And, of course, you may be able to intercept packets that contain information revealing the owners and operators.

Here in San Francisco, where I have logged nearly 200 APs and Wi-Fi cards, no way. If I were to hire a squad of war drivers and walkers and spend hundreds of hours researching, I might get details on some but probably never all of them.

Some of them are so far away that seeking them out would be an exhaustive task—I can get a weak signal from The Embassy Hotel, which is seven blocks away and has a dozen tall buildings in the signal path. An organization called BAWRN (Bay Area Wireless Research Network) has a public access AP that is even farther, 12 blocks or easily a mile. If I steer the antenna carefully, I can just barely detect this AP. And as with The Embassy and virtually all of the other APs, the signal is far too weak to actually connect to them—I could never associate with them and use them for Internet access. (Which incidentally is likely to be illegal, depending on the AP and the people who own it).

See the chapter on war driving for more.

Q: If someone is using CommView to monitor my AP, is there any way I can detect it?

A: Not according to the producer. CommView puts the card in stealth mode so it doesn't normally transmit any signal.

Q: I am seeing hundreds of MAC addresses showing up when I scan. There can't be that many wireless computers within range, can there?

A: Maybe. Using CommView in Search mode, I have found a single AP on a given channel (by carefully positioning the grid antenna) that has a dozen or more icons—the little desktop computer without the "rabbit-ears," meaning that they are not APs. Some of these single APs have names that indicate they are small home wireless networks, such as "joeathome," and which are unlikely to have more than a couple computers in that home network. You may be seeing individual computers that are part of this small network, perhaps wired using CAT-5 type cable. But when the numbers seem unlikely, then what you are seeing may be "ghosts"—images that may be generated by signals mixing with each other. In other words, you are seeing MAC numbers of imaginary machines; they don't exist.

There is little need to be concerned, other than just curiosity, wanting to know, as you aren't likely to intercept any data from them, only what goes through the AP.

Also, remember that you may be seeing a wireless card, rather than an AP, that is in ad-hoc mode. If the card has an external antenna, then others may well be able to monitor the card unless you are in RMM stealth mode.

Also, some wireless card drivers generate "garbage packets" that are part of real packets that have become, somehow, put together, or include parts of damaged packets.

Q: How can I determine the physical location of an AP that doesn't transmit any information that would reveal this?

A: To begin, first look at any intercepted packets and see what this tells you. Analyze every bit of data you intercept. If you snarf e-mail messages, what do they tell you about the person that sent them? Analyze. Make logical conclusions.

If you don't get any useful info from the content of the packets, such as the Web sites you see that they are visiting, then use the SSID and do a little detective work. Start with Google or do a Copernic search. It might show up in the many lists of public access hot spots, which are open to the public for free or fee. But don't limit your search to just that. For example, there is a public AP located near here that has a rather unusual SSID—the first names of the two people who set it up. Now, I knew from that fact that I can receive their signal and because I am almost surrounded by tall apartment buildings, that it has to be within a few blocks.

The first name is one I had never heard before; let us call him "jonzo."

Now by carefully moving the grid antenna (which you will remember is very directional) I think I have managed to rule out reflected signals (multipath), and because of the angle, it probably has to be in one of the taller buildings. So I put on my messenger dispatcher hat and call the building managers of several apartments in what I believe to be the signal path. "I'm calling from TreeFrog messenger service. We have a package to deliver to someone but part of the address label was water damaged. Do you have someone named "jonzo" in your building?"

After half a dozen tries, I hit pay dirt. "Oh, yeah, he's here. You can drop the thing off at the manager's office and we'll see that he gets it."

The SSID might be an acronym for a business. JONAIL might be Jo's Nails, a manicure place. And what you don't find on Google you just might find in the trusty Yellow Pages.

Q: Is higher power always better?

A: No. Most of the time, higher power is not only unnecessary but is also undesirable. You only need enough power to make a good connection, and 30mW is often enough for home use. Higher power means your signal goes out farther and can thus interfere with another system or be eavesdropped by someone else.

Q: How can I adjust the power level?

A: With most cards, you cannot. Read the Help files and e-mail the card manufacturer (who just might reply) and the Help files for the software you are using. If it cannot be adjusted, you can do so with the ERP. Sort of.

Q: ERP? What the hell is that?

A: Effective radiated power. It is the technique of concentrating a radio signal, of "focusing" it into a narrower beam; making it more directional. If you have more than one directional antenna, use the one with the lowest dB gain that still can make a solid connection with the AP you want to monitor.

Q: What is a peer-to-peer wireless network?

A: This is a network where two computers with wireless cards (PCI or PCMCIA) connect directly to each other. To establish communications, each user enters information from the other computer's card. This is the "opposite" of infrastructure, where the cards work through an access point.

Q: I have seen, here and elsewhere, mention of "layers." What is this and what does it mean?

A: First, it is not necessary to understand this to use everything published here. But since I have been asked, here goes.

There is a seven-layer model in transmitting (sending and receiving) data in a network that uses TCP/IP technology, which is what the Internet uses as well as some private networks. The layers are, from bottom to top, physical, data link, network, transport, session, presentation, application. Should you want to memorize, then remember the phrase, "Please Do Not Throw Sausage Pizza Away." Information, in the form of packets or datagrams, is processed through these layers where it goes from raw data into the packets, and has a header attached that includes the destination (IP) address and other things.

Q: What is Chipset?

A: You see this frequently when reading about PCMCIA Wi-Fi cards. It refers to the type of chips used in the card. The two most often used are Prism (Senao card and others) and Hermes, used in some Orinoco cards. A good place to get current information is Seattle Wireless, http://www.linux-wlan.org/docs/wlan_adapters.html.gz.

Part Six

REAL SPIES, REEL SPIES, AND REAL "NUTZ"

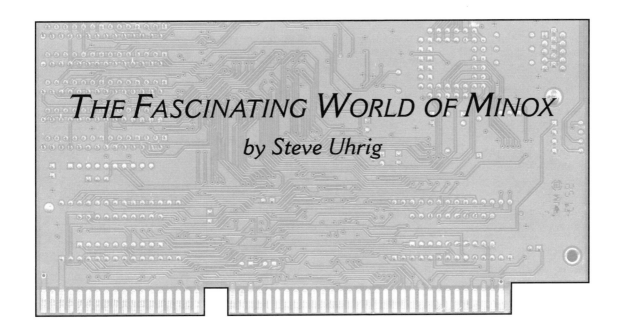

THE FASCINATING WORLD OF MINOX
by Steve Uhrig

MINOX. THE SPY CAMERAS MADE FAMOUS IN MOVIES AS WELL AS REAL LIFE.

The various James Bonds have used Minox cameras, as have royalty such as King Frederik of Denmark and Prince Philip, Duke of Edinburgh, who has a gold one. Just Plain Folks own thousands of these submini cameras as have spies from the year the Minox was developed to the present. Soviet spy John Walker, for example, used the electronic Model C Minox.

The inventor, Walter Zapp, wanted to create a camera that was small, yet able to take excellent quality photographs and use film much smaller than 35 mm. The first Minox was the Riga, named for the city in Latvia, and it was manufactured from the late 1930s to the mid-'40s.

Then came the IIIs, which is the smallest mechanical Minox. It has no means to measure light levels, so you need to estimate light levels and the proper shutter speed to use it. The model after the IIIs was the model B. The mechanics basically are the same between the two; however, the B has a light meter added onto one end. The B was the most popular model, produced in large quantities. Therefore they are readily available for rather low prices. Most Minoxes you see offered for sale will be model B.

An improvement on the B was the model BL, which is very similar to the model B mechanically. Many consider the BL to be the best pocket camera. It has an accurate light meter usable indoors or outdoors, a rugged mechanical design, and still is fairly small. The BL was produced on a limited scale, so they are scarce and expensive.

Following the BL was the model C. This was the first electronic Minox, which means it has a photocell and an electronic circuit to read the light level and automatically set the shutter speed. The shutter also is electrically operated rather than mechanically as in all earlier Minoxes. Many prefer this, and believe electronics are more accurate, repeatable, and reliable than mechanics. It also is the longest of all the submini Minoxes. It is still pocket sized though. The C was a common model and is an accurate, easy-to-use camera. The automatic shutter seems to read your mind and give perfect exposures even when you'd swear it couldn't because of complicated lighting or whatever. You can override the automatic shutter if you care to for whatever reason, up to 1/1000 second. Or you can leave it in automatic as most do. I highly recommend the Minox C as a starter camera. It is accurate, easy to use, inexpensive, and reliable.

The next step up from the model C is the Model LX. This is a more recent electronic camera and is the current design. The main difference between it and the other models is a faster shutter (up to 1/2000 where 1/1000 is the fastest speed for any of the other metal Minoxes). This will let you catch faster action and use faster film outdoors where it would be overexposed in any other camera. The LX is smaller than the C, larger than the B or BL, and a joy to see or use. It is fairly expensive, relatively speaking, because it is a new camera.

Then you have the plastic Minoxes, the model EC and the currently offered ECX. These are the smallest Minoxes, about the size of a Bic cigarette lighter. They are electronic, all automatic, and have no user settings; they are point-and-shoot cameras.

The EC and ECX are low-cost little brothers to the classic metal Minox. Their advantages are that they are inexpensive, small, rugged, very simple to use, and capable of superb pictures. Their disadvantages are that there are no adjustments the user can override, and the lens is one stop slower than any other Minox, meaning you need twice as much light. Also on the downside, the maximum shutter speed of 1/500 means that you may overexpose pictures taken outside with fast film, and accessories for all other Minoxes do not fit the EC. The closest distance they can focus is 3 feet, where other Minoxes can focus down to 8 inches. The EC and ECX have fixed-focus lenses and an electronic-only shutter.

The EC/ECX are fine little cameras to carry with you always; they are so small as to nearly disappear in your pocket along with your keys and change. The minimal metal content means they do not trip airport metal detectors, making them attractive to travelers. They are cheap, so if you lose one you aren't out a week's pay.

Now, there is no way of knowing how many spies have used Minox over the last half century, but when you consider how widespread espionage

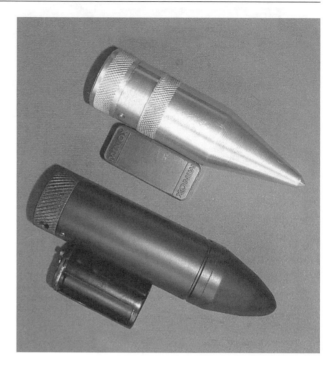

These "bullets," called dead letter drops, were used to hold the Minox film. They could be pushed into the ground someplace where an operative would later pick them up.

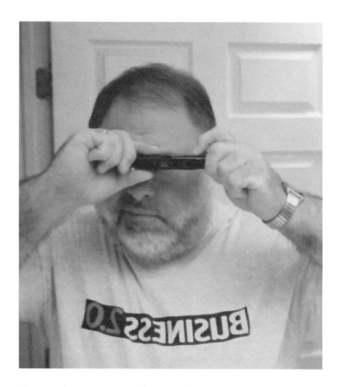

Steve takes a picture of himself.

is, no doubt there have been many thousands. Walker, for example, wore out three Minox model Cs in photographing the estimated one million documents he stole and gave to the Soviets.

Steve Uhrig is a Minox expert and owns one of the largest collections of these high-quality cameras in the world. You can see his inventory at http://www.swssec.com/minox.

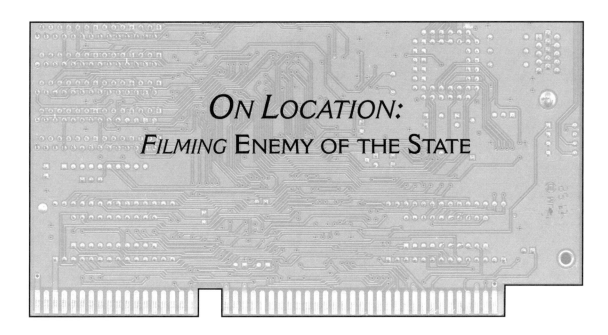

On Location: Filming Enemy of the State

Electronic surveillance, spying, in the movies and on TV has little resemblance to how it is in the real world. OK, it is true that today, Maxwell Smart's shoe phone would actually work, what with phones getting smaller and more obnoxious, but back when that TV series was produced, it was not possible. The technology just didn't exist. And remember when Uncle Joe of *Petticoat Junction* overheard a visitor speaking into a fountain pen saying, "Rooshin spy calling Moscow"?

The Man From U.N.C.L.E. and *Mission: Impossible* perpetuated the myth that a transmitter the size of a lemon drop could send a signal not just above the chimney tops but halfway around the world.

Enemy of the State is an exception to the typical unrealistic treatment of the subject of spying because Disney-Touchstone hired surveillance expert Steve Uhrig as a consultant. They even found a role for him as the owner of a spy shop. The photos below were taken on location. To see more photos, visit this book's Web site at http://www.fusionsites.com/dbm2.

Steve in front of his dressing room on the movie set.

This surveillance helicopter, used in the movie, is equipped with two FLIR systems and SWS' S band air-to-ground wireless video link (along with some other systems we can't discuss).

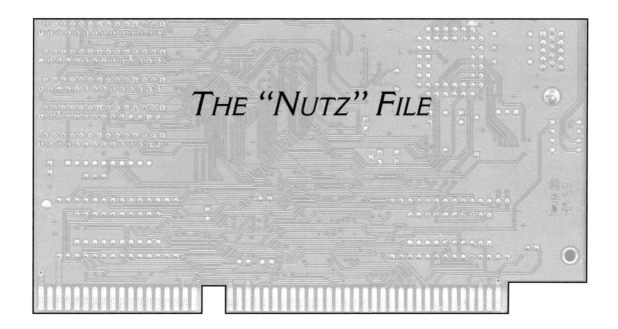

The "Nutz" File

THIS CHAPTER TAKES ITS NAME FROM THE JARGON OF THE SURVEILLANCE industry, from the technicians and engineers who are part of the inner circle. It consists of true stories of people who believed they were under surveillance, but weren't. Or were they?

However, the word "nutz" is not intended as an insult or a put-down; it is just a term that somehow came into use and stuck. Based on experience, on descriptions made by these unfortunate people, and on the technology that we have learned over the years, we know that in most cases what these people describe just are not within the realm of what is technically possible in electronics. We try to explain this to them, to make them understand that what they found and are convinced is a listening device just isn't.

For example, Betsy's Bug in *The Bug Book*. What I found in her phone was a voltage regulator used back in the early days of having a fax machine on the same line as the telephone. Another example, from the same book, was a filter used to keep AM radio stations from interfering with the phone line.

And in *Don't Bug Me* is the story of the employee who believed that a bug was in a glass of orange juice that his employer had given him. He was actually "examined" using a complete suite of countermeasures equipment. With negative results.

And then, again, it is always possible that some of them actually *are* under surveillance, and the fact that they are nervous, upset, and apparently having emotional problems does not rule this out.

Coming up are a few examples from Steve's files. You may notice that some of the writers use actual technical surveillance terms mixed with nonsense words and phrases. They are reproduced here as received except for names or other identifying information, which has been sanitized.

Authors' note: This chapter is not by any means intended to ridicule the people who have submitted these letters. To them, the surveillance they believe in is very real. However, we in the business know that with few exceptions this reported spying is not real. This is based partly on having a technical perspective (what is and is not possible) and partly on having developed the intuition that is one ingredient of a good countermeasures technician.

These letters are reprinted as received except that we have corrected some (but not all) of the spelling errors. Having read most of this book by now, you may see where some of the "technical" terms don't make sense. So, as another exercise in learning, you might read them carefully and see which of these terms I refer to. I'll give you one example from the first letter:

I just want to know if I am hearing harmonics? Two tone harmonics that have two ranges. High and low. Is it possible that I am hearing transmitters or bugs or signals.

Hearing transmitters or bugs? Not aurally. With a receiver, yes, if there were a transmitter.

What makes sense? What seems logical and electronically possible?

Also, please be aware that Steve does not normally do countermeasures sweeps for individuals in their homes, and that the cost of any sweep is well in excess of the $1,000 mentioned below.

Hello,
I do not claim to know anything about surveillance equipment and nor did I want to know. But my neighbors are into high tech team work on new science equipment and I have been the victim of their trials which include telescope CCD camera's and distance ranging scanning equipment and some other stuff I wouldn't even begin to try and explain.
I have only one question that would stop me from continuing to feel insane.
Please help.
I just want to know if I am hearing harmonics? Two tone harmonics that have two ranges. High and low. Is it possible that I am hearing transmitters or bugs or signals.
I know this sounds amateurish. But I have been suffering from their exploits for too long and I need to find out what I am dealing with.
Please tell me if I am hearing signals that sound like they are coming from somewhere in space but I hear it in my house and car. Nowhere else. Been to the doctor and had an MRI. This is not something that I could make up in my head nor is it tinitus (ear ringing)
Vivian

****

Dear Mr. Uhrig;
Several years ago, you sent me an email about a post that I made on ___ _____'s surveilance list. I had said that I had a sweep of my house and you were concerned that the sweep was not made by someone who was qualified.
You were, in fact, correct. The sweep was done by a local private detective agency who later told me that they had specifically purchased the RF detector to do my sweep. Naturally I was quite annoyed when I heard this.
I have since found out who the culprit is. I just can't prove it. I live in a townhouse in Baltimore County, next door to a Russian engineer. Because one finally does come to know when they are being spied on, I realized finally that it was this Russian engineer who was listening to everything in my house.
About three years ago, I came to realize that he is also looking into my house with some kind of pinhole camera and/or fiber optic. It is not a peeping tom kind of thing. He is not acting alone. There is some politics involved and some unscrupulous lawyers.
What's happening to me is so unfair. I can't seem to get help from law enforcement. Because I am constantly getting hints from some people who some lawyer is trying to sue in some kind of malpractice suit, I have actually learned some things. I have a condition called obsessive compulsive disorder and apparently some one was evesdropping on me and the lawyer thought I might have witnessed something. The truth is, I'm in the middle of something that I know nothing about.
Can you imagine walking around in your own house and being under total surveilance (just like Will Smith in Enemy of the State) *Only in my case, it is being done through the walls. Also, I'm positive that there is some kind of camera planted in several of my ceiling fans. One of the biggest problems is the cathedral ceilings. I'm positive that he is looking into my hallway. But the ceiling is so high that even the*

person who did drywall work could not reach it.

Trust me Mr. Uhrig, my story is almost as interesting as Enemy of the State.

I get followed wherever I go. Things that I say and do end up as phrases in the Baltimore Sunpaper. I'm in a no win situation. My husband and I bothwork on state jobs. We are not politicians. Just middle class working people. However, I have worked for almost 20 years for the State Department of _____ and _____ at 301 West _____ St. There is no shortage of politics there. And as I said because of this, I have not been able to get any real help with my privacy problem.

If I could just get someone like you to believe me, I would be so happy. I'm so tired of people telling me that I'm crazy. I refuse to move out of my townhouse.

The funny thing is, I never suspected that my next door neighbor was the "SPY". But after the sweep yeilded nothing and all of those strange things kept happening, I realized that it was him that was doing this to me.

Also, every time a private investigator came to my house, he made sure that his house was guarded. Now I'm absolutely positive he is the one who has been doing this to me. If you can give me any advice about how to get my privacy back, I would be most appreciative. I really believe my story is one of human interest.

The Susan Wilson camera story is nothing compared to what has happened to me. And yet, I can't prove it or get anyone to investigate it because of the politics.

Thanks for listening. Hope to hear from you.
Happy Easter!
L

****

To whom it may concern,

I am employed by a defense contractor. I have been receiving spam with cryptic messages in the email aliases that reference personal or other information about me, topics discussed in telephone conversations on my home phone and even reference to websites I have viewed at the workplace a government site.

Though plausibly deniable, I have made a link between the emails, someone with access to activities at the government site, and person(s) who have my home telephone line tapped. A certain government agency had previously tried to recruit me in the D.C. area and seemingly were attempting to do so again this past summer. Since that time I have been receiving this spam. I have made official attempts with the government contractor to get to the bottom of the situation. In November I was arrested and I believe the arrest was a set-up. Over this last weekend I switched to a new clean hard drive and switched ISP's. I was up and running for all of Friday evening.

On Saturday my ports 443 and 25 - Secure Sockets Layers and Outbound SMTP was shut down on both the new ISP accounts on a clean hard drive and my previous ISP account on my old hard drive. I checked with both the ISP and my Internet hosting service upstream on these difficulties and neither indicated there was a problem.

My other hard drive is in a horrible state as I think there is some spyware on it as I have a huge number of ports open. As far as the old hard drive goes I think someone broke into my residence and gained physical access to the drive. Back to the spam. It had been previously suggested I could communicate with this entity (I cannot be sure who it is - yes this worries me) through forms on spam sites. Mainly I have just submitted annoying messages up until recently. Recently I asked them what they want. I received a reply back in a spam alias.

What is most alarming is that in order to get my ports 443 and 25 turned on I actually asked this entity by submitting the request into a spam mail form. I know this stuff sounds implausible but I have established a pattern over the last few months and have managed to "call them out" for one by switching ISPs and keeping my head up. If it is indeed a government agency I only wonder what they want. However, the total lack of privacy—Internet, Telephone and possibly bugs in my home I find difficult to tolerate. So the things I do know—I am being monitored by someone telephonically, my Internet is controlled and I would think monitored upstream, my ports have been shut down when I switched to a new ISP on both my old and the new ISP (so there is a device upstream), Harassed (see previous) and lastly I believe I was set-up and arrested. Plausible deniability is high in all of this but there is definitely a pattern and to the best of my knowledge I would say that laws are being broken, I am being targeted by someone AND my privacy is being flagrantly and overtly violated. If you would like to see this interesting email form I can provide it. It is hosted in China but the form submits to a U.S. IP address.

I am usually a pretty tolerant guy but this stuff in really getting on my nerves

They are aware I am sure that I am on to the surveillance

As far as I know this email account has not been compromised.

Any suggestions

XXX

****

Dear Sir,

I am a 55 year old ex-Marine Viet Nam, college graduate, loan officer. I have been back to work for nearly a year now. In 1997 I started hearing voices. The voices continue to this day. I know that it is people doing this to me with some kind of device implanted in my head and or ears possibly attached to the auditory nerve. I had a small device that looked like a coiled up snake dug out of me in the Nam in 1967. I am sure these people are pumping RF into me and out of me via this device. I have lost my career, real estate, reputation and nearly my sanity due to this situation. No I am not crazy. I am lucid and able to function as a normal person.

I need some help though. I know that RF is detectable and that someone such as yourself must know what this is that is affecting me. Perhaps you can direct me to someone who can locate the device and or disable it. I am not wealthy but I can and will pay someone who can tell me the truth and/or assist me in ending this nightmare with the voices.

Respectfully and Sincerely Yours,

D. E.

****

I can hear voices through my heat/air vents. I live in a mobile home. I can hear my boyfriend talking and he can hear me (he says I'm crazy) from his barn, and I think from his home.

I believe from his home, because I can hear his wife talk from time to time. He also has something in my car, he seems to know where I am always at. And he has said things to me that he could only hear if he was here, or listening.

What can I buy to find this out!

I also want to trace it back to him, he is very wealthy, and I know he has violated my civil rights. I live in a very small town, where money talks, and unless I could prove it, know one would ever believe it.

I can also hear music at time, similar to a cell phone ring. At times I can call his barn and hear my phone ringing, and hear it if I leave a message.

Please help me!!

Thanks

S

****

Dear Steve,

I have experienced many problems in my phone line since a long time ago. I hired a company to check my home, he did not find any bug at the time that he inspected my home. However he concluded in his report that my telephone was tapped.

I suspect about my neighbor. I assume that company did not find any bug the day that he inspected because my neighbor has a camera in his window pointing the entrance of my home, so he knows who come in or out from my home. The day of the inspection he disconnected all the proves that company could find. The following are the signs that I have in my home:

* I have noticed strange sound or volume changes on my phone line. I have popping or scratching sounds coming from my phone handset when it's hungup.

* My phone rings and nobody is there.

*I have high pitched squeal/beep in my line.

* My radio has suddenly developed strange interference.

*My TV develops strange interferences. It turned off many times without touch the remote control. My TV has "Pops" even when it is unplugged.

*When I watch TV until late a green sign "Sleep Time" shows up. I did not set up my TV with this feature.

* I have many sounds like "click" "click" in all the bathrooms in my home (I have 3).

I have some questions: It is possible for you to

catch the bugs even when they are disconnected? How much will cost me? if you find something when you finish your job are you going to give me a written report? I find you because the company who inspected my home used TSCM products, I opened in the Internet and your are one of the Recommended U.S. TSCM firms.

Thanks,
S

****

Dear Mr. ____,
Sorry for not respond your e-mail before, my daughter got married and I was running for everyplace.

Thank you very much for your response and I would like to make some comments about Mr. ____, he really is a big layer. The day that he come my home after he "did all the tests", he grand son asked me my phone number and I gave the number to him. Mr. ____ told me Ah! that is your number, of course I said, I think he was testing my neighbor's phone number.

His grand son after he heard my telephone he said to me that looks that my line is connected to a computer. After Mr. ____ finished the job in my home he offered me to get a lawyer, I was waiting almost 3 months at the end he said to me that nobody answered him.

He asked me to get photos from my neighbor to compare the telephone hooks with my telephone hook, what I did. When he got the photos he went to my office and he offered me to get a lawyer to see my case, but when I get the money we are going to split half and half. As I do not have time and I though that maybe he knew someone I said yes, however after 3 months that he did not get anybody he wanted to talk to me. We met in McDonalds near to my home. He came to ask more money for the phone calls that I made to him. I said, why you are charging me for the phone calls I called you because I wanted to know if you get the lawyer that you offered to me.

Well he said that it time consumed. I showed him the e-mail that he sent to me in which he said that he is going to give a writing report after 5 days that he finished the job. I asked for the report and he gave after 4 months that he finished the job. When he saw his e-mail he did not ask money any more.

Also he said to me that he knew someone who works for the government and he can get a copy of my neighbor's computer, so I can have a prove that my neighbor is spying me and my family, I asked about this several times his last response was "I can't get him, maybe he retired"

He got for me a form to send the papers to an Agency that see this cases in ____ County. I am complaining about the telephone company, I sent his report, the pictures that I had, my neighbor's hooks pictures. They responded me that the telephone company has 3 months to contact me about the case, if the telephone company does not do anything in three months I have to inform them and they will force to do something to the telephone company.

In August the 8th. will be the 3 months.

Mr. ____ charged me 2,500.00 for his job. I am going to fax you his report as soon as I can. I do not have computer I have to come to the Library. You are recommending me a private detective, what I would like to have is a prove that my neighbor is spying me and my family, I am a 100% sure that he is doing it, but I need a prove. Do you think that a private detective can get that information? please advise me. How much approximately will cost me?

Again, thank you very much for your response,
S

****

Some time ago, a company in the United Kingdom advertised a device that would, they claimed ... well here is the blurb, as was posted on an Internet newsgroup:

Hi, Ive been reading the post on GPS tracking and thought I would introduce a non GPS/GSM or even battery/mains powered tracking system with a range of 600Km. The crucial, most important feature is that the system is totally passive.

There is absolutely no R.F. present. This has been tested with the best equipment on the market. Furthermore the tracking device or bug as we call it does not operate by battery power. The size of the bug is extremely small and they are made in different shapes, the largest one is like a vitamin capsule. The smallest one is to be implanted inside a tooth of the object.

The operational range of the tracking system is guaranteed to 6000 kilometres.

Looking at the technologies on the market today, the signal from the ALPHA FIVE MIST) can not be jammed in any way

The absolutely main factor here is that the bug is 100% passive. The suspect will never know that there is anything installed or placed. More strategically on covert operations because no equipment will give any indication that an operation is going on. We even can supply the bugs in metal or plastic versions. Metal detectors, or non-linear junction detectors will of course not detect the latter.

The truly "invisible" bug is here.

Nothing else on the market today has these specifications.

more info at http://www.spy-equipment.co.uk/Tracking/tracking.html

G

Someone answered the post:

Wow, I can't believe that they're still hawking this crap.

I had my hands on this unit over 10 years ago at a surveillance trade show. It's nothing more than a block of plastic with a telescopic "rabbit ears" t.v. antenna and an insertable card tray that held what looked like carbon paper. I was told by the huckster that the trays (purchased separately of course) held the molecular composition of whatever you were looking for, be it bugs, narcotics, explosives etc.

The antenna was on a swivel base and would "point" to the target. You just followed it like a Divining Rod. All that was necessary was to insert the card of that particular "molecular property" to make the "Detector" sensitive to that item. You would purchase cards for each type of eavesdropping device, narcotic or explosive.

Today they've jumped on the bandwagon offering Bio-chemical and GPS cards. Prime Time or 20/20 did an expose' story on this company where they had conned some police dept. into buying these units to sniff out drugs in highschool lockers, cars, persons etc. Needless to say that it failed miserably.

This company was successfully prosecuted by the Federal Govt. and paid some major fines. This product is SO outrageous that even _____ won't touch it. Now THAT tells you something.

K

The original poster retorted:

I am amazed at people like you who have nothing better to do than slander products which you know nothing about! This product is real and available with guaranteed tracking distance of 6000 Km we have demonstration units in stock and can prove its ability should you wish to put your money were your mouth is. If you cant get your head around it that's fine I understand after all you don't know everything!!

G

****

Steve:

I live in the DC area and got your name from the _____ site (which was an excellent site). I was hoping I can schedule you for a limited sweep. I was inquiring to see if you can do a limited house sweep for possible bugs/microphones in my house possibly in my car. I understand from your website the charge is $1000 for that service. I would pay more if it requires it but I hope this would be sufficient. I have no problem paying for that if it resolves my problem or you think its worth the effort.

I know it seems ridiculous and am still in denial this is happening but I believe I am bugged by my immediate neighbors. I cant prove it and I am not crazy or paranoid. I wish it was that simple. I don't know their motivation except to say they appear to have an intense dislike of me and it seems they would be most happy if I move.

I'd be most happy if they were deported and maybe I can if you find anything since they're violating a federal law and their resident aliens they could be in violation of their green cards—they are _____s

I will soon be a nutcase that if this is not resolved because this situation is ruining my life. I just want to see if you can find any device that they may have in or around the house and in my automobiles. I am very serious about paying for your services if you can help and I am no quack. I have an engineering degree myself (no electronic bkgrd unfortunately) and work for the _____ Office in _____ City Virginia. If you do find any bug great but if you do not I will be moving out of my house shortly because I cant live with this nonsense any longer.

Can I talk to you to schedule a limited sweep of my home and cars (only 2)? I appreciate your time. Thank you.

P. M.

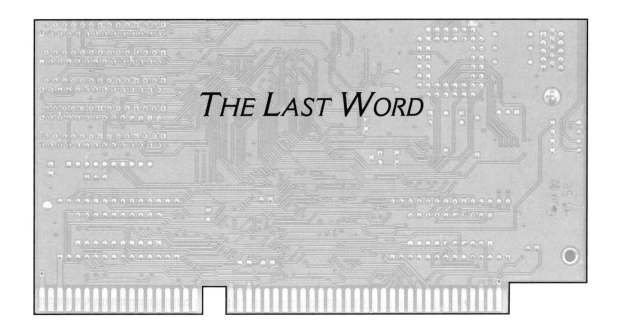

THE LAST WORD

HAS TECHNOLOGY GONE TOO FAR?

I remember when a satellite malfunctioned and pagers all over the San Francisco area stopped working. And other things stopped working. People stopped working because they weren't able to communicate. I was running errands that day, and while I didn't see anyone freaking out, I heard a lot of nervous comments on the street.

If a single cell site in any large city were to shut down for a few days, the other cells (to which traffic reverts) would be so overloaded that tens of thousands of phones would not work. Some phones would be locked out, as there is a part of the system that allows preference to signals used by law enforcement and emergency services and, naturally, the federal government. People have become so dependent on cell phones and pagers, they cannot function without them.

Dilbertville.

An executive at Cisco Systems prophesized that eventually most everything could be networked, connected together—fireplaces and Venetian blinds and toasters and Mr. Coffee and vacuum cleaners. And what with more and more control over our lives, we may one day live in the futuristic society described by Ray Bradbury in *There Will Come Soft Rains*, where every aspect of life is automated.

"In the living room the voice-clock sang, *Tick-tock, seven o'clock, time to get up, time to get up, seven o'clock!*"

And if the government privacy invaders have their way, this networking would include your radios and TVs. Signals could be sent that turn them on to a particular channel for the latest propaganda.

"The wedding of sophisticated information-gathering techniques with computerized information storage and dissemination systems has created, for the first time, a very real danger that the sense of privacy which has traditionally insulated Americans against the fear of state encroachment will be destroyed and be replaced, instead, by a pervasive sense of being watched. The emergence of such a police state mentality could mean the destruction of our libertarian heritage."

—Source unknown

I thought I had coined a new phrase, "Techno-Luddite," but the term has been around for some time. Nevertheless, we all have one thing more or less in common, and that is that life can get too technical and that it is possible to be surrounded and controlled by gadgets. Gimmicks.

Now, this may seem strange coming from someone who has all the stuff that you see in this book, so I will rationalize: The equipment was necessary to write this book. But now that the manuscript has been sent to the publisher, much of it has been sold.

Meanwhile, the Techno-Luddites are a fast-growing group of people who want to become less of a slave to the beeping gadgets that society requires them to tote around everywhere they go.

"Luddism and the Neo-Luddite Reaction. Cultural change necessarily involves resistance to change. The term Luddite has been resurrected from a previous era to describe one who distrusts or fears the inevitable changes brought about by new technology. The original Luddite revolt occurred in 1811, an action against the English Textile factories that displaced craftsmen in favor of machines. Today's Luddites continue to raise moral and ethical arguments against the excesses of modern technology to the extent that our inventions and our technical systems have evolved to control us rather than to serve us and to the extent that such leviathans can threaten our essential humanity."

—Martin Ryder
University of Colorado at Denver
School of Education
http://carbon.cudenver.edu/
~mryder/itc_data/luddite.html

There must be limits upon what the government can know about each of its citizens. Each time we give up a bit of information about ourselves to the government, we give up some of our freedom. For the more the government or any institution knows about us, the more power it has over us. When the government knows all of our secrets, we stand naked before official power, stripped of our privacy. We lose our rights and privileges. The Bill of Rights then becomes just so many words.

—Senator Sam Ervin

Now, not many people are going to give up pagers and cell phones just because of what is written in this book. But they will at least be in a better position to safeguard their privacy.

People will continue to depend upon computers and more and more home wireless

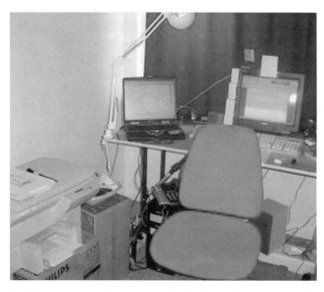
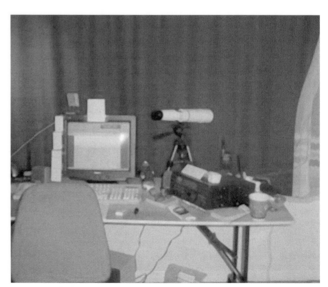

This is my new work area. Much of the equipment seen elsewhere in this book has been sold.

The Last Word

networks will be set up as it becomes more necessary to have more than one PC and because of the freedom to move around that Wi-Fi offers.

But having read this, you are well on your way to making wireless networks secure.

Broadband over powerline may not turn out to be as widespread as the inventors hope, but if it becomes available in your area and the salespeople come knocking on your door, you will know how to deal with it. And them.

So the industries will come up with new and better ways, technology, to invade your privacy and control your lives.

And writers, hackers, geeks, and others who love the right to be let alone will fight back with books and gadgets and ideas.

And the game goes on …

—M. L. Shannon
November 2004
Wellington, New Zealand

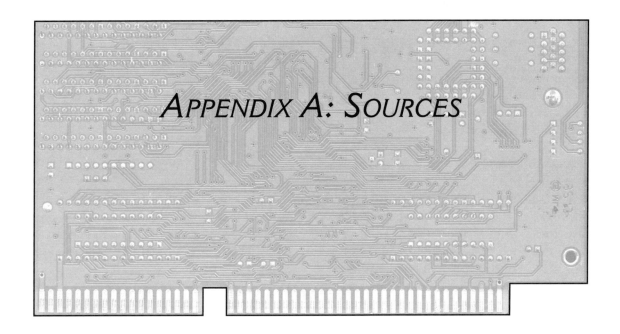

Appendix A: Sources

COMPUTER PARTS

COMPUTER PARTS ARE WHERE YOU FIND THEM; SOME CITIES HAVE QUITE A few independent retailers, some do not. But don't overlook **Circuit City**, as they just might have what you want. The people are great to deal with and their return policy is, well, better than some of the other big nationwide chains.

TELEPHONE ACCESSORIES

Shomer-Tec is an old, established company that I have done business with over a period of years. They are honest, reliable, and sell good quality products.

Their catalog (the 2004 edition is 95 pages) lists all sorts of telephone accessories including automatic tape recorder starters, wrong-number generators, and legal (depending on its use) infinity transmitters. There are several long-play tape recorders, digital recorders, extension monitors, and a wide variety of law enforcement products, some available to the public and some restricted.

Five bucks for this catalog is well spent just for the information it contains. Find them online at http://www.shomertec.com.

For information on the laws governing recording of audio conversations, visit the Web site of the **Reporters Committee for Freedom of the Press** at http://www.rcfp.org/taping/.

The Seeker wireless network detector was originally sold by Chrysalis Development, LLC. at http://www.chrysalisdev.com/.

Last time I checked the link no longer worked but the same device, called **QuickLink Mobile Wi-Fi Seeker**, is available from http://www.smithmicro.com.

CABLES AND CONNECTORS

Back in the introduction to wireless networking, I rambled on about how you could waste a lot of money buying things that won't work with other things. The good news is that there is a small company in Maryland where they know about cables and connectors; this is their specialty and they have been at it for 25 years.

The RF Connection may have in stock exactly what you need, or they will custom make it. This company comes highly recommended by the coauthor of this book, and his word is gold. Check out their site. You'll find what you need at http://users.erols.com/rfc/index1.htm.

PC CARDS

The **AbsoluteValue Systems** Web site has the most complete list of wireless cards and specs I know of. Take a look at http://www.linux-wlan.org/docs/wlan_adapters.html.gz.

If you want a Senao PC card, look for a **Surf and Sip** in your area. For locations, see their Web site at http://www.surfandsip.com/.

ANTENNAS

Try **NetNimble Wireless Products** at http://www.netnimble.net. I have purchased several antennas from them, including the 24-dB grid pictured in the wireless networking chapters. While their prices are not the lowest on everything, they have a decent selection, ship promptly, and take COD orders. However, I do not recommend that you buy cables or pigtails here; two cables I bought were defective.

WEB SITES

There are a gazillion sites that have information on what this book is about, so I will list only a few. Check them on a regular basis and you can stay on top of what's new in wireless networking and the Internet.

Slashdot's motto is "News for Nerds. Stuff that matters." Visit them at http://slashdot.org/.

WhatREALLYhappened.com is mostly politics but often has related articles on computing you won't find in the mainstream media. Go to http://whatreallyhappened.com/.

Also check out the **Electronic Frontier Foundation** at http://www.eff.org/ and the **Electronic Privacy Information Center** at http://www.epic.org/, and don't forget *Electronic Surveillance and Wireless Network Hacking* on the Web at http://www.fusionsites.com/DBM2. Updates to various chapters will be posted periodically, and I may set up a private chat area if there is sufficient interest.

Appendix B: Suggested Reading

Wiretap Detection Techniques: A Guide to Checking Telephone Lines. Theodore N. Swift. Thomas Investigative Publications, Inc., 1997.

This book can be ordered from http://www.angelfire.com/biz/investigator/index.html.

A Guidebook for the Beginning Sweeper. Glenn H. Whidden. Technical Services Agency.

Do not be deceived by the title; this book is not just for beginners. Whidden's book will be useful as a thought-provoking reference work for anyone who engages in sweeping or engages sweepers (counter-eavesdropping specialists). It is not a technical manual but instead deals with tactics that can defeat a resourceful and highly professional eavesdropper. Other things such as a sweeper's fee structure, legal considerations, and a glossary of terms are included.

There is also a discussion of the eavesdropper's problems and the vulnerabilities that he will display and how he will try to reduce them.

Published by Technical Services Agency, it is unfortunately out of print, so it is where you can find it.

Network Bondage. Steve Gibson. Gibson Research Corporation: http://grc.com/su-bondage.htm

This can be found on the ShieldsUP! Web site, which calls itself "the Internet's quickest, most popular, reliable and trusted, free Internet security checkup and information service." It's one of a series of articles about Windows' lack of security and what you can do about it. Somewhat technical, but well written.

Hacker Proof: The Ultimate Guide to Network Security. Lars Klander, Edward J. Renehan Jr. Jamsa Press, 1997.

An older book, but an excellent work on understanding the Internet, TCP/IP protocols, and, as the title states, how to make a network very secure.

The Fugitive Game: Online with Kevin Mitnick. Jonathan Littman. Little, Brown, 1997.

The story of Kevin Mitnick while on the run and his eventual capture.

The Cuckoo's Egg: Tracking a Spy Through the Maze of Computer Espionage. By Clifford Stoll. Pocket, 2000.

This is one of the most fascinating books on hacking I have ever read, one of the early cases of tracking a hacker. And hacking a tracker!

PRIVACY AND SURVEILLANCE CLASSICS

In law school, first-year students spend a great deal of time learning about landmark high-court decisions to help them understand how the law works. The following titles are old but highly recommended for anyone who wants to see surveillance, privacy, and government intrusion as it has progressed over the years. These books illustrate the foundation of what we have today.

Cheesebox. Gerard Michael Callahan, with Paul Meskil. Prentice-Hall, 1974.

Autobiography of Gerry "Cheesebox" Callahan, one of the greatest wiremen of his time. Fascinating book! Out of print, but worth searching for. Reminds me of many things I cannot talk about.

The Rise of the Computer State. David Burnham. Random House, 1983.

Stranger on the Line: The Secret History of Phone Tapping. Patrick Fitzgerald. Vintage/Ebury, 1987.

Methods of Electronic Audio Surveillance. David A. Pollock. Charles C. Thomas Pub. Ltd., 1972.

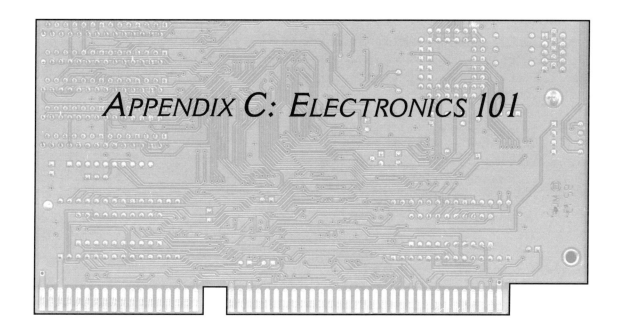

Appendix C: Electronics 101

Having some knowledge of basic electronics is useful for working in countermeasures. There may be times when you will need to use a meter—either an analog volt-ohm-meter (VOM) or the digital counterpart, a digital multimeter (DMM)—to check out a wire discovered in an office that cannot be explained that could be part of a past—or present—surveillance device. Or to test the telephone line voltage.

Ohms law is an uncomplicated formula in DC electronics using E for voltage, measured in volts; I for current, measured in amps; and R for resistance, measured in ohms. It can be expressed three ways:

$$E = I \text{ times } R$$
$$I = E \text{ divided by } R$$
$$R = E \text{ divided by } I$$

Now, to illustrate this, picture a water tower with a pipe coming out the bottom, and a tap that can be opened or closed.

Voltage is represented by gravity that makes the water want to flow through the pipe. Voltage is also called electromotive force or potential difference.

Current is represented by the flow of water through the pipe.

Resistance opposes this flow and is represented by the size of the pipe and the tap that controls the flow.

Gravity (voltage) wants to cause the water (current) to flow through the pipe (wire) so if you increase the amount of water in the tank, there is more pressure, so more water (current) will flow, depending on the size of the pipe (wire) and how far you open the valve (resistance).

The larger the pipe, the more water can flow, if the tap is open. Adjusting the tap controls how much water (current) flows through the pipe (wire). Closing the tap stops water flow completely, known as an "open circuit."

AC electronics is more complicated, with terms such as impedance (sort of like AC resistance), reluctance, inductance, and so on, but you won't need to get into all that.

Just being able to use a meter to check out suspicious wires is a good start. But be careful because you don't know what (how much voltage, if any) might be there.

Also keep in mind that whenever an electrical current flows through a wire, it produces heat. In the chapter on thermal imaging, you saw pictures made from the heat generated by electronic spy devices.

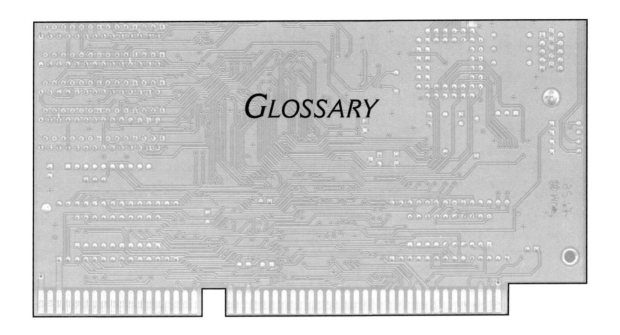

GLOSSARY

NETWORKING TERMS

10Base-2—An Ethernet wiring standard that uses bus topology, thin coaxial cable, has a maximum length of 185 meters, and runs at 10 Mbps.

100Base-T—A wiring standard that works like 10Base-T but can run at 100 Mbps. 100Base-T is also known as "fast Ethernet."

10Base-T—The most common Ethernet wiring standard. 10Base-T uses twisted pair wiring that's used to connect buildings' telephone wire to the telephone company, runs at 10 Mbps, uses a star network topology, and is limited to a length of 100 meters.

802.11—Wireless networking standards developed by the Institute of Electrical and Electronic Engineers (IEEE). The three currently in use are 802.11a, 802.11b, and 802.11g.

802.11a operates in the 5 GHz band; is about five times faster than 802.11b; has less interference because the frequency band is less crowded; and is not compatible with 802.11b or g.

802.11b operates on 2.4 GHz, as do microwave ovens and some wireless telephones; and sends and receives at 11 Mbps.

802.11g operates on 2.4 GHz, as do microwave ovens and some wireless telephones; is about 5 times faster than 802.11b; and works with 802.11b.

802.1x—A wireless encryption method, supposedly more secure than WEP. I haven't tried it yet.

Access Point (AP)—A device that client adapters (wireless cards) use to connect to a wired network. The wired network can be as simple as a single desktop computer with a DSL modem wired to a router that has the access point built in, like my network with the Siemens router described here.

Ad-Hoc Mode—The method used to create a wireless network between two or more computers without the need for an access point. They communicate directly with each other.

Bluetooth—A wireless networking technology that works over a short distance, 20 feet or so, used mainly as a cable replacement. Wireless headphones, mice, keyboards, etc., can use Bluetooth.

Client Adapter—One of the two components needed to set up a wireless network. A device with which a computer—portable or desktop—can connect to (associate with) an access point. A PCI or PCMCIA card.

DHCP—Dynamic Host Configuration Protocol, a protocol by which a server automatically assigns IP addresses to clients so network administrators don't have to configure them manually.

Hot Spot—Wireless APs that are found in public places such as airports, convention centers, hotels, and coffee shops. Also, people's private networks that they have left wide open to hackers because they didn't read this book.

Infrastructure Mode—The usual way of creating a wireless network in which clients associate with, communicate with, or connect to an access point.

IP Address—An Internet Protocol (IP) address is a unique numeric identifier for a computer or device on a network differentiating it from all other devices. Just like telephones have their own unique number.

LAN—A local-area network (LAN) is a computer network that extends over a small area. It's usually within one building but can go between different locations.

MAC Address—A unique identifier permanently burned into devices on a network. A NIC, Network Interface Card, for example, has a "permanent" MAC. It is possible, under some circumstances, to alter or "spoof" this number.

Promiscuous Mode—The mode of a packet sniffer in which it intercepts all traffic on a LAN, or segment of a LAN. It is both a maintenance and a hacking tool, depending on who is using it and why.

Roaming—Similar to cell phones, a system where users can automatically be connected to a different AP if they move around in the facility, such as from one building to another. Cisco makes such a system and has a nice animated demo you can download.

Router—Simplified explanation: A device that "routes" network traffic and gets it where it is supposed to go by analyzing the destination address (IP) in the packets it receives. It uses look-up tables to determine the route. Essentially the same as a gateway, sort of.

SSID—An SSID, or service set identifier, is the name used by a wireless access point to identify itself. Also may be called the "network name." The SSID can be anything the owner wants it to be.

SSL—Secure sockets layer (SSL) is a way of encrypting information, such as credit card numbers. How secure it is depends upon who is trying to make it un-secure.

Seven-Layer Model—A description of the processes data goes through to be transmitted using TCP/IP protocols. Described somewhere in the text. I forget where. This is a big book. You don't expect me to remember every little detail, do you?

Sniffer—A device, software or hardware, that searches for wireless networks. Also, software used to intercept and analyze data on wired networks. See also *Promiscuous Mode*.

Subnetting—You don't wanna know. I'm serious. It is a complicated process that I had to learn for my CompTIA Network+ Certification, in which computers are assigned their unique IP so that other devices can find them to communicate. DHCP does this for you, but remember that if you set up a network, you can select which IPs are allowed access, such as only your own computers. This keeps hackers out. Maybe.

TCP/IP—Except for "connectionless" protocols such as TFTP (trivial file transfer protocol) and another unknown one that certain people on the SF2600 IRC channel use, TCP/IP are the basic protocols for communications between computers on a network. Oversimplification: TCP, transmission control protocol, is what puts data packets together in the order they were sent, or transmitted.

Topology—Any of several configurations through which computers in a wired network are connected together. Ring, Star, Mesh, and Bus are the most common, though Bus isn't used much any more. Ring and Star are just that. Mesh has a connection from every device to every other device.

Expensive as it requires a lot of cables, but the most secure; it means that no matter what happens to one or more hosts, such as crashing, all the others will still be connected.

WEP (Wired Equivalent Privacy)—A feature used to encrypt and decrypt data signals transmitted between WLAN devices.

Wi-Fi—Wi-Fi, or wireless fidelity, is the common term for a wireless local-area network (WLAN). The term is used generically when referring to any type of 802.11 network, whether 802.11b, 802.11a, dual-band, etc.

Wireless LAN—You already know this. Why are you looking here?

WIRELESS TELEPHONE TERMS

AMPS—Advanced Mobile Phone Service.

Base Station—A transmission and reception station for handling cellular traffic. Also referred to as a cell site.

CDMA—Code division multiple access. A system, which includes some cell phones, that uses spread spectrum for both voice and data.

Cell—A site. It is the basic unit of a cellular system, from whence comes the name. A given area is divided into small cells, each of which has a radio transmitter and receiver. The cells vary in size depending on terrain and population and number of users.

Control Channel—A channel used for transmission of digital control information in Manchester (conventional analog phones) from a cell site or a cell phone.

ESN—Each cellular phone is assigned a unique ESN, or electronic serial number, which is automatically transmitted to the cellular base station every time a call is placed. The MTSO (mobile telephone switching office) validates the ESN with each call. Cloned cellular phones transmit a stolen ESN and charges are made to the real cellular phone account.

Frequency Reuse—The cellular system has a limited number of available frequencies. So the original system was set up with the cells arranged to use a certain number of the frequencies in a way that no adjacent cells use the same ones. That way they don't interfere with each other.

GSM—Global system for mobile communications.

GPRS—General packet radio service, a data transmission system used on GSM phone systems, similar to cellular packet data.

Handoff—The process by which the MTSO passes a cellular phone conversation from one frequency in one cell to (usually) another frequency in another cell. The handoff is performed so quickly that users usually never notice.

Hertz—The basic unit of frequency, formerly known as "cycles." An alternating current starts at zero volts (the baseline), builds to its positive peak (110 volts for example) then back to zero, reverses to the negative peak and then back again to the baseline. This is one cycle. Applies also to time varying DC.

MTSO—Mobile telephone switching office. The CO (central office), or "switch," that processes calls to and from cell phones.

NAM—Number assignment module. A chip that stores information in a cell phone.

PCS—Personal communication services. Similar to cellular, but uses digital.

Registration—The procedure that a cellular phone uses when connecting to a base station to indicate that it is now active.

SID—System identification. A five-digit number that indicates which service area the phone is in. Most carriers have one SID assigned to their service area.

TDMA—Time division multiple access. Multiplexing. TDMA takes many signals (phone calls for example) and "cuts" them into small slices of time, milliseconds, so that they can all transmit at the same time.

About the Authors

M.L. SHANNON

Producing books about surveillance was not originally my idea.

In one of my electronics courses in college, we were all required to write an essay and read it to the class. Well, what I wrote was about spying, and the true story behind how I made this decision, the reasons, you can read about in *Don't Bug Me*. Since I am into throwing puzzles and exercises at my readers, to get them to think and analyze and learn, I won't specify which story it was, but with a little detective work, you'll figure it out. If you have read *Don't Bug Me*, that is.

So, anyway, after class, the instructor asked me why I didn't write a book on the subject. I didn't give it much thought at the time, but while we were partying after the dreaded finals week, and talking about the things we had experienced that year—including the incident I just hinted at—somehow it came up, as it involved someone else in that class. My lab partner, Jo Anne, said she hoped I would, that she'd look forward to seeing it published. And so, some years later, I wrote *Don't Bug Me*.

Much has happened since *Don't Bug Me* was published, and while I haven't had the adventures that my coauthor Steve Uhrig has, there have been some interesting times. I went to work for a small electronics company that made spy equipment, and I loved working there where, being the only tech, I had a lot of freedom. And I got to make a number of trips to Mexico where I delivered, tested, set up, and maintained some of our spy equipment for the Federales. Real spy stuff, but my Spanish is so poor I understood very little of what was being intercepted. I most definitely did not need to know.

Just before I returned to the United States, I was invited to go on a drug raid. I was given a bullet-resistant vest and ordered to stay near the car and not approach the house that was the objective. I was really excited about that, but, unfortunately, the raid was cancelled.

Disappointed, yes, but I had a great time at the hotel and got to fly to Tucson in a Beechcraft King Air. Beautiful airplane. Wish I could have had the right seat!

All good things come to an end: it turned out that the boss was selling some of the devices we made to the wrong people. I was very fortunate that I was on another trip to Mexico, delivering some cellular radio interception equipment, when the company was raided by Customs and the Secret Service.

A few months later, I went to work as a sweep technician for one of the few truly professional countermeasure companies, where we searched various and sundry corporate offices for listening devices, and I had the honor to be invited to be a guest speaker in Burbank at TREXPO, the Tactical Response Exposition, which was a convention of law enforcement officers. SWAT teams. The "black Velcro" people. I was allowed into the exhibition hall where there were enough automatic weapons to start World War Six.

Since then, I have written *The Bug Book* and *The Phone Book*, both published by Paladin Press, and a number of magazine articles, including "Cyber-Street Survival," published in *Nuts & Volts* magazine. This six-part series on Internet privacy and security are the basis of one chapter in this book.

And, recently, I have had several articles published in *Blacklisted! 411* magazine.

In all of this I have gained experience and knowledge, which contributed to my latest work.

STEVE UHRIG

Steve Uhrig is the founder and president of SWS Security, a multinational manufacturer of electronic surveillance and intelligence-gathering systems for government agencies. His company provides electronic intelligence systems and support services to law enforcement, government, and military entities worldwide handling such operations as drug interdiction, antiterrorism, and military intelligence and counterintelligence.

Steve at the Secret Service shooting range.

Uhrig's interest in electronics began during early childhood. He attributes his aptitude to his father, who encouraged his interest in the field and taught him basic electronic theory and practical application by helping him build hobby electronic projects and repair electrical items for neighbors.

One of Uhrig's lifelong hobbies has been ham radio, where he holds an advanced class amateur radio license WA3SWS. He is also a member of several fraternal and professional organizations and is a Fellow in the Radio Club of America, a member of the Armed Forces Communications and Electronics Association, the Association of Former Intelligence Officers, an instructor for the National Technical Investigators Association, a founding member and former director of the international Minox Historical Society, and MENSA.

Uhrig is a published author with more than 338 technical articles and White Papers on electronic surveillance and intelligence operations to his credit. Papers he has written are required reading for students in certain government intelligence agencies. Uhrig also consults extensively to the news and entertainment media and has appeared on more than 200 talk shows, news interviews, and discussion panels pertaining to surveillance, privacy, and related topics.

In the late 1990s, Uhrig was contracted as a filmmaker, hired as surveillance consultant to the blockbuster Hollywood technothriller *Enemy of the State*. Uhrig worked for nine months writing portions of the script and training actors, including stars Gene Hackman and Will Smith, in surveillance equipment installation and operation. He was on set daily during filming as a consultant to producer Jerry Bruckheimer and director Tony Scott. In addition, Uhrig had a small acting part in the movie, writing his own lines in a scene with Hackman and Smith as an electronics store merchant selling underground surveillance equipment.

Uhrig formed his own company in 1972 doing small surveillance and communications projects for local police departments. In the mid-1970s he entered federal government service as a full-time pursuit and worked for a host of U.S. government agencies in various electronic support operations. While in government service, Uhrig worked heavily in the field with military agencies from many countries and developed a reputation for being capable and reliable, an artisan who could perform the impossible under extreme conditions.

In the mid-1980s, Uhrig left government service to devote full attention to his growing company.

Since that time his firm has expanded to include facilities in 10 countries and support services to more than 40 foreign governments. He has developed electronic surveillance and intelligence-gathering systems that have become the worldwide standard in the fight against terrorism, as well as state-of-the-art defense systems created to protect both people and property, which also are in use around the world. Uhrig's advancements in surveillance technology in such areas as video transmission technology, electronic tracking and direction finding, communications and signal intercept, and clandestine communications have established performance standards throughout the industry.